RECOGNISING ACHIEVEMENT

G C S E
Mathematics

Graduated Assessment

Stages 5 & 6

Authors

Howard Baxter

Mike Handbury

John Jeskins

Jean Matthews

Mark Patmore

Contributor

Colin White

Series editor *Brian Seager*

Hodder & Stoughton
A MEMBER OF THE HODDER HEADLINE GROUP

Orders: please contact Bookpoint Ltd, 130 Milton Park, Abingdon, Oxon OX14 4SB.
Telephone: (44) 01235 827720, Fax: (44) 01235 400454. Lines are open from
9.00 – 6.00, Monday to Saturday, with a 24 hour message answering service.
Email address: orders@bookpoint.co.uk

British Library Cataloguing in Publication Data

A catalogue record for this title is available from The British Library.

ISBN 0 340 801905

First published 2001

Impression number 10 9 8 7 6 5

Year 2007 2006 2005 2004 2003 2002

Cover illustration by Mike Stones.

Produced by Gecko Limited, Bicester, Oxon.

Printed in Italy for Hodder & Stoughton Educational, a division of Hodder Headline
Plc, 338 Euston Road, London NW1 3BH by Canale.

Acknowledgements

The Publishers would like to thank the following individuals and companies for
permission to reproduce photographs in this book:

Life File Photo Library: Mike Evans page 57.
Robert Harding Picture Library: page 215.
The Photographers Library: page 126, 181, 212.

Every effort has been made to trace ownership of copyright. The Publishers would
be happy to make arrangements with any copyright holder whom it has not been
possible to trace.

This book covers the last part of the specification for the Foundation tier of GCSE Mathematics and also the first part for the Intermediate tier. It is particularly aimed at OCR Mathematics C (Graduated Assessment) but could be used for other GCSE Mathematics examinations.

The work in this book covers the criteria in stages M5 and M6, and aims to make the best of your performance in the module tests and the terminal examination:

- Each chapter is presented in a style intended to help you understand the mathematics, with straightforward explanations and worked examples.
- At the start of each chapter is a list of what you should already know before you begin.
- There are plenty of exercises for you to work through and practise the skills.
- At the end of each chapter there is a list of key ideas.
- After every four or five chapters there is a revision exercise.
- Some exercises are designed to be done without a calculator so that you can practise for the non-calculator sections of the papers.
- Many chapters contain Activities to help you develop the necessary skills to undertake coursework.
- At frequent intervals throughout the book there are exam tips, where the experienced examiners who have written this book offer advice and tips to improve your examination performance.
- Revision exercises and Module tests are provided in the Teacher's Resource.

Part of the examination is a calculator-free zone. You will have to do the first section of each paper without a calculator and the questions are designed appropriately.

The percentage of the marks for the Assessment Objectives on the module tests and terminal examination are:

- 10% AO1 Using and Applying Mathematics
- 40% AO2 Number and Algebra
- 20% AO3 Shape, Space and Measures
- 10% AO4 Handling Data

The remaining marks to balance AO1 (10%) and AO4 (10%) are awarded to the internal assessment (coursework).

Most of the marks given for Algebra in AO2 are for 'manipulative' algebra. This includes simplifying algebraic expressions, factorising, solving equations and changing formulae. Some questions are also being set which offer you little help to get started. These are called 'unstructured' or 'multi-step' questions. Instead of the question having several parts, each of which leads to the next, you have to work out the necessary steps to find the answer. There will be examples of this kind of question in the revision tests and past examination papers.

Top ten tips

Here are some general tips from the examiners to help you do well in your tests and examination.

Practise:

1 all aspects of **manipulative algebra** in the specification
2 answering questions **without** a calculator
3 answering questions which require **explanations**
4 answering **unstructured** questions
5 **accurate** drawing and construction
6 answering questions which **need a calculator**, trying to use it efficiently
7 **checking answers**, especially for reasonable size and degree of accuracy
8 making your work **concise** and well laid out
9 using the **formula sheet** before the examination
10 **rounding** numbers, but only at the appropriate stage.

Coursework

The GCSE Mathematics examinations will assess your ability to use your mathematics on longer problems than those normally found on timed written examination papers. Assessment of this type of work will account for 20% of your final mark. It will involve two tasks, each taking about three hours. One task will be an investigation, the other a statistics task.

Each type of task has its own mark scheme in which marks are awarded in three categories or 'strands'. The titles of these strands give you clues about the important aspects of this work.

For the investigation tasks the strands are:

● Making and monitoring decisions – what you are going to do and how you will do it
● Communicating mathematically – explaining and showing exactly what you have done
● Developing the skills of mathematical reasoning – using mathematics to analyse and prove your results.

The table below gives some idea of what you will have to do and show. Look at this table whenever you are doing some extended work and try to include what it suggests you do.

Mark	Making and monitoring decisions	Communicating mathematically	Developing the skills of mathematical reasoning
1	organising work, producing information and checking results	discussing work using symbols and diagrams	finding examples that match a general statement
2	beginning to plan work, choosing your methods	giving reasons for choice of presentation of results and information	searching for a pattern using at least three results
3	finding out necessary information and checking it	showing understanding of the task by using words, symbols, diagrams	explaining reasoning and making a statement about the results found

Mark	Making and monitoring decisions	Communicating mathematically	Developing the skills of mathematical reasoning
4	simplifying the task by breaking it down into smaller stages	explaining what the words, symbols and diagrams show	testing generalisations by checking further cases
5	introducing new questions leading to a fuller solution	justifying the means of presentation	justifying solutions explaining why the results occur
6	using a range of techniques and reflecting on lines of enquiry and methods used	using symbolisation consistently	explaining generalisations and making further progress with the task
7	analysing lines of approach and giving detailed reasons for choices	using symbols and language to produce a convincing and reasoned argument	report includes mathematical justifications and explanations of the solutions to the problem
8	exploring extensively an unfamiliar context or area of mathematics and applying a range of appropriate mathematical techniques to solve a complex task	using mathematical language and symbols efficiently in presenting a concise reasoned argument	providing a mathematically rigorous justification or proof of the solution considering the conditions under which it remains valid

For the statistical tasks the strands are:

- Specifying the problem and planning – choosing or defining a problem and outlining the approach to be followed
- Collecting, processing and representing data – explaining and showing what you have done
- Interpreting and discussing results – using mathematical and statistical knowledge and techniques to analyse, evaluate and interpret your results and findings.

The marks obtained from each task are added together to give a total out of 48.

The table below gives some idea of what you will have to do and show. Look at this table whenever you are doing some extended work and try to include what it suggests you do.

Mark	Specifying the problem and planning	Collecting, processing and representing data	Interpreting and discussing results
1–2	choosing a simple problem and outlining a plan	collecting some data; presenting information, calculations and results	making comments on the data and results
3–4	choosing a problem which allows you to use simple statistics and plan the collection of data	collecting data and then processing it using appropriate calculations involving appropriate techniques; explaining what the words, symbols and diagrams show	explaining and interpreting the graphs and calculations and any patterns in the data

Mark	Specifying the problem and planning	Collecting, processing and representing data	Interpreting and discussing results
5–6	considering a more complex problem and using a range of techniques and reflecting on the method used	collecting data in a form that ensures they can be used; explaining statistical meaning through the consistent use of accurate statistics and giving a reason for the choice of presentation; explaining features selected	commenting on, justifying and explaining results and calculations; commenting on the methods used
7–8	analysing the approach and giving reasons for the methods used; using a range of appropriate statistical techniques to solve the problem	using language and statistical concepts effectively in presenting a convincing reasoned argument; using an appropriate range of diagrams to summarise the data and show how variables are related	correctly summarising and interpreting graphs and calculations and making correct and detailed inferences from the data; appreciating the significance of results obtained and, where relevant, allowing for the nature and size of the sample and any possible bias; evaluating the effectiveness of the overall strategy and recognising limitations of the work done, making suggestions for improvement

Advice

Starting a task

Ask yourself:
- what does the task tell me?
- what does it ask me?
- what can I do to get started?
- what equipment and materials do I need?

Working on the task

- Make sure you explain your method and present your results as clearly as possible
- Break the task down into stages. For example in 'How many squares on a chessboard', begin by looking at 1 × 1 squares then 2 × 2 squares, then 3 × 3 squares. In a task asking for the design of a container, start with cuboids then nets, surface area, prisms ... Or in statistics you might want to start with a pilot survey or questionnaire.
- Write down questions that occur to you, for example, *what happens if you change the size of a rectangle systematically?* They may help you find out more about the work. In a statistical task you might wish to include different age groups or widen the type of data.

- Explore as many aspects of the task as possible.
- Develop the task into new situations and explore these thoroughly.
  - What connections are possible?
  - Is there a result to help me?
  - Is there a pattern?
  - Can the problem be changed? If so, how?

Explain your work

- Use appropriate words and suitable tables, diagrams, graphs, calculations.
- Link as much of your work together as possible, explaining, for example, why you chose the tables and charts you used and rejected others, or why the median is more appropriate than the mean in a particular statistical analysis, or why a pie chart is not appropriate. Don't just include diagrams to show identical information in different ways.
- Use algebra or symbols to give clear and efficient explanations; in investigations, you must use algebra to progress beyond about 4 marks. You will get more credit for writing $T = 5N + 1$ than for writing 'the total is five times the pattern number, plus one'.
- Don't waffle or use irrelevant mathematics; present results and conclusions clearly.

State your findings

- Show how patterns have been used and test conclusions.
- State general results in words and explain what they mean.
- Write formulae and explain how they have been found from the situations explored.
- Prove the results using efficient mathematical methods.
- Develop new results from previous work and use clear reasoning to prove conclusions.
- Make sure your reasoning is accurate and draws upon the evidence you've presented.
- Show findings in clear, relevant diagrams.
- Check you've answered the question or hypothesis.

Review/conclusion/extension

- Is the solution acceptable?
- Can the task be extended?
- What can be learned from it?

Example task

On the next page there is a short investigative task for you to try, in both 'structured' and 'unstructured' form. The structured form shows the style of a question that might appear on a timed written paper. The unstructured form represents the usual style of a coursework task. The structured form leads you to an algebraic conclusion. Notice the appearance of algebra from question 4 onwards, through a series of structured questions. These mirror the sort of questions you would be expected to think of (and answer) if you were trying it as coursework.

Comments about the questions, linking the two forms of presentation, are also shown.

Although the task in both forms directs you to investigate trapezium numbers, you would be expected to extend the investigation into other forms of number, such as pentagon numbers, to achieve the higher marks.

ACTIVITY

structured form

Trapezium numbers

These diagrams represent the first three trapezium numbers.

Each diagram always starts with two dots on the top row.

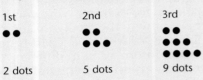

1st   2nd   3rd

2 dots   5 dots   9 dots

So the third trapezium number is 9 because nine dots can be arranged as a trapezium. There are two dots in the top row, three dots in the next row and four dots in the bottom row.

1 Write down the next two trapezium numbers

2 **a)** Draw a table, graph or chart of all the trapezium numbers, from the first to the tenth.
     b) Work out the eleventh trapezium number.

3 The 19th trapezium number is 209. Explain how you could work out the 20th trapezium number without drawing any diagrams.

4 Find an expression for the number of dots in the bottom row of the nth trapezium number.
Test your expression for a suitable value of n.

5 Find, giving an explanation, an expression for the number of dots in the bottom row of the diagram for the $(n + 1)$th trapezium number.

6 The nth trapezium number is x. Write down an expression in terms of x and n for the $(n + 1)$th trapezium number. Test your expression for a suitable value of n.

unstructured form

Trapezium numbers

These diagrams represent the first three trapezium numbers.

Each diagram starts with two dots on the top row.

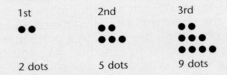

1st   2nd   3rd

2 dots   5 dots   9 dots

So the third trapezium number is 9 because nine dots can be arranged as a trapezium.

Investigate trapezium numbers

NB Although the task in this form asks you to investigate trapezium numbers, you have the freedom to – and are expected to – extend the investigation to consider other forms of number such as pentagon numbers.

Commentary

This question allows you to show understanding of the task, systematically obtaining information which **could** enable you to find an expression for trapezium numbers.

This question provides a structure, using symbols, words and diagrams, from which you should be able to derive an expression from either a table or a graph. Part **b)** could be done as a 'predict and test'.

In the unstructured form you would not normally answer a question like this.

From here you are **directed** in the structured task, and **expected** in the unstructured task, to use algebra, testing the expression – the **generalisation**.

In the unstructured form this would represent the sort of 'new' question you might ask, to lead to a further solution and to demonstrate symbolic presentation and the ability to relate the work to the physical structure, rather than doing all the analysis from a table of values.

Stage 5

1 Cuboids

You should already know

- how to use simple scales
- how to use a calculator to multiply decimals.

Isometric drawings

For these drawings, you will need a ruler, a pencil and triangle spotty paper.

Isometric drawing is a way of representing three-dimensional shapes. Triangle spotty paper is ideal for this.

Exam tip

Make sure that the spotty paper is the right way round. Draw in pencil, then any mistake can be put right.

This is the wrong way round.

EXAMPLE 1

Look at this shape.

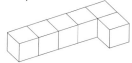

This is an isometric drawing of the same shape.

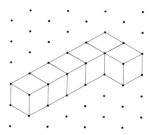

You can make the drawing look more realistic and easier to understand by using different shadings for the sides facing in the three different directions.

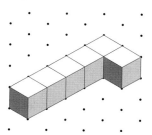

EXAMPLE 2

On triangle spotty paper make an isometric drawing of each of these shapes.

Answers

a)

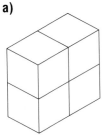

b)

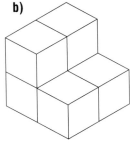

a)

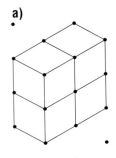

b)

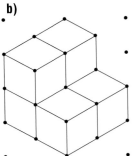

EXERCISE 1.1A

1 On triangle spotty paper, make isometric drawings of these shapes.

a)

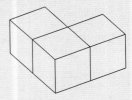

b)

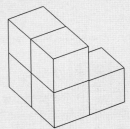

2 On triangle spotty paper make isometric drawings of
  a) A cube 2 by 2 by 2
  b) A cuboid with dimensions 4 by 3 by 2.

3 On triangle spotty paper make isometric drawings of **a)** and **b)**.

a)

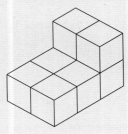

b)

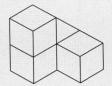

Exercise 1.1A cont'd

4 How many cubes make up this shape?

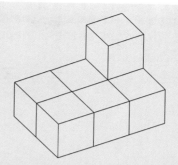

5 Use triangle spotty paper to make isometric drawings of these solid shapes. Use a scale of 1 centimetre to 2 units.

a)

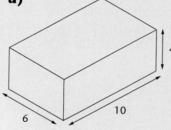

4

6

10

b)

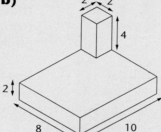

2 ⌐ 2

4

2

8

10

c)

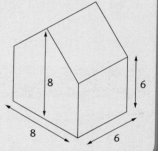

8

6

8

6

EXERCISE 1.1B

1 On triangle spotty paper make isometric drawings of these shapes.

a)

b)

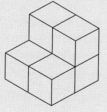

2 This shape is made from five cubes. Draw all the other different shapes that can be made from just five cubes in one layer. (**Hint**: There are eleven more to find.)

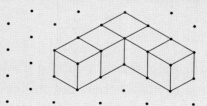

Exercise 1.1B cont'd

3 Use triangle spotty paper to make isometric drawings of these shapes.

a) Use scale 1 cm to 1 m

3 m 2 m 4 m 2 m

b) Use scale 1 cm to 5 m

5 m 10 m 20 m 25 m 10 m

c) Use scale 1 cm to 6 cm.

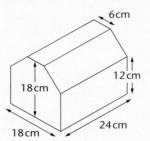

6 cm 18 cm 12 cm 18 cm 24 cm

4 Find a simple cupboard at home and measure its dimensions. On triangle spotty paper make an isometric drawing of the cupboard to a suitable scale.

Volume of a cuboid

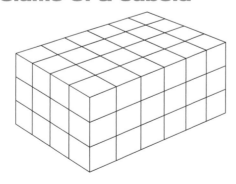

> **ACTIVITY 1**
>
> If you have some multilink cubes, make this cuboid and find its volume.

The diagram shows a cuboid made up of centimetre cubes.

On the top layer there are six rows of four cubes, so there are 6 × 4 = 24 centimetres cubes in the top layer.

There are three layers. So in all there are 3 × 24 = 72 cubes.

This means that the volume of the cuboid = 72 cubic centimetres or 72 cm^3.

The calculation for this was $6 \times 4 \times 3 = 72$, so the formula for the volume of a cuboid is:

Volume of a cuboid = V = length × width × height or $V = lwh$.

ACTIVITY 2

Make some more cubes and cuboids with multilink cubes and check the formula works.

Exam tip

A cube with side 1 cm is called a centimetre cube and is written cm^3.
Make sure that the length, width and height are all in the same units.
The units of volume are then the cube of those units e.g. cm^3 or m^3.

EXAMPLE 3

Calculate the volume of a cuboid with length 8·5 cm, width 6·4 cm and height 3·6 cm.
Volume = $8.5 \times 6.4 \times 3.6 = 195.84 \, cm^3$.

EXAMPLE 4

For the cuboid in the diagram, calculate the volume
Volume = $6 \times 3 \times 2 = 36 \, cm^3$

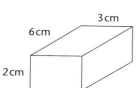

EXAMPLE 5

A concrete path is laid which is 20 m long, 1·5 m wide and 10 cm thick. Calculate the volume of concrete used.

Thickness = 10 cm which needs to be changed to metres.

Thickness = $10 \div 100 = 0.1 \, m$

Volume   = $20 \times 1.5 \times 0.1 = 3 \, m^3$

EXERCISE 1.2A

1 Find the volumes of these cuboids, all lengths are in centimetres.

a)

b)

2 A box has height 10 cm, length 5 cm, width 3 cm. Find its volume.

3 A classroom is 6 m long, 4 m wide and 3 m high. Work out its volume.

4 A biscuit tin is 12 cm long, 5 cm wide and 6 cm deep. Work out its volume.

5 A cube has edges 5 cm long. Find the volume of the cube.

6 A shoe box has a base 10 cm by 15 cm and is 30 cm deep. Calculate the volume of the box.

7 A cuboid has a base 3 cm by 4 cm and its volume is 48 cm^3. What is the height of the cuboid?

8 A cuboid has a volume of 48 cm^2. Find the dimensions of three different cuboids with this volume.

EXERCISE 1.2B

1 Find the volumes of these cuboids, all lengths are in centimetres.

a)

b)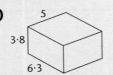

2 A box has height 12 cm, length 5·3 cm, width 3·8 cm. Find its volume.

3 A classroom is 6·7 m long, 4·3 m wide and 2·8 m high. Work out its volume.

4 A box of chocolates is a cuboid. It is 12·5 cm long, 6 cm wide and 4·6 cm deep. Work out its volume.

5 A cube has edges 7·4 cm long. Find the volume of the cube.

6 A fish tank is 15 cm wide, 60 cm long and 25 cm deep. Find the volume of the tank.

7 A pane of glass is 150 mm wide, 600 mm long and 5 mm thick. Find the volume of the pane of glass.

8 A cuboid has a volume of 200 mm^3. Find the dimensions of three different cuboids that have this volume.

Nets of cuboids

The net of a cube or cuboid is the flat shape that will fold to make the cube or cuboid.

EXAMPLE 6

Draw the net of this cuboid. All the sides are in centimetres.

This net is not drawn to size but the lengths are marked.

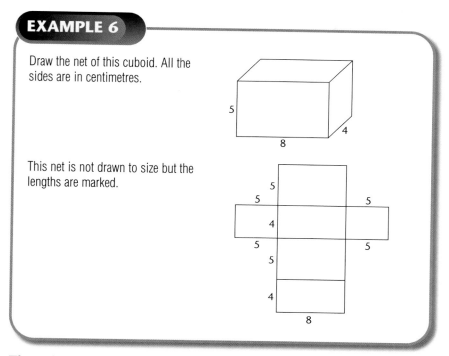

The edges that meet together are the same length.

Try drawing this net, cutting it out and folding together to make the cuboid.

The sides that are fixed on either side of the main rectangle can be placed in other positions.

Try and see which will work. Use squared paper to help you.

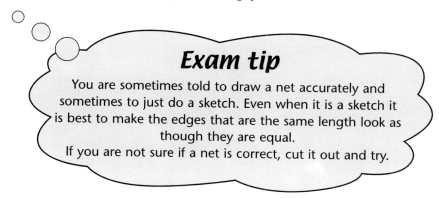

Exam tip

You are sometimes told to draw a net accurately and sometimes to just do a sketch. Even when it is a sketch it is best to make the edges that are the same length look as though they are equal.

If you are not sure if a net is correct, cut it out and try.

As a cuboid has six faces, the net needs six rectangles but not every arrangement of the six rectangles will fold together to make the cuboid.

If you are asked to draw the net of a box with no lid then the net will be made up of five rectangles.

EXAMPLE 7

Which of these nets will fold to make a cuboid?

They are drawn on squared paper to size.

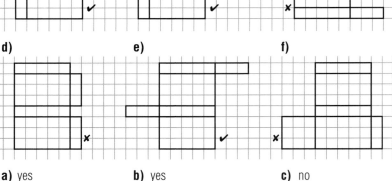

a) yes   **b)** yes   **c)** no

d) no   **e)** yes   **f)** no.

EXERCISE 1.3A

1 Draw a net for each of these cuboids on squared paper. All lengths are in centimetres.

a) 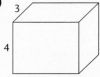   **b)**   **c)**

Exercise 1.3A cont'd

d)   e) 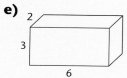 f)

2 Which of the nets below are for a cuboid? They are drawn accurately on squared paper.

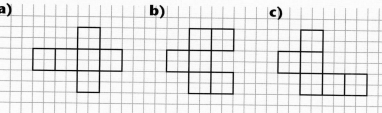

a) b) c)

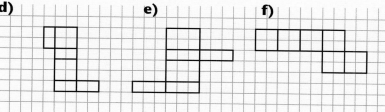

d) e) f)

3 A cuboid packing case has a base 40 cm by 40 cm and a height of 80 cm. It has a top. Draw the net for the case. Use a scale of 1 cm to 20 cm.

4 A dice is a cube with 2 cm edges. The numbers on the faces are 1, 2, 3, 4, 5, 6
The two numbers on the opposite faces add up to 7.
Draw the net of the dice and write the numbers on the faces.

5 A box containing paper has a base 15 cm by 20 cm and is 5 cm high.
It has no top.
Draw the net for the box. Use a scale of 1 cm to 5 cm.

EXERCISE 1.3B

1 Draw a net for each of these cuboids on squared paper. All lengths are in centimetres.

a)

b)

c)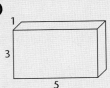

d)

e)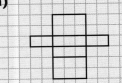

f)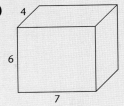

2 Which of the nets below are for a cuboid? They are drawn accurately on squared paper.

a)

b)

c)

d)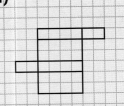

e)

f)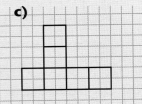

Exercise 1.3B cont'd

3 The box containing my radio is 20 cm by 10 cm by 5 cm.

It has a top.

Draw the net of the box on squared paper. Use a scale of 1cm to 5 cm.

4

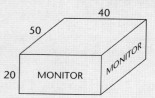

This is the box for my computer monitor.

It says MONITOR on each of the sides.

Draw the net for the box on squared paper. It has a lid. Use a scale of 1 cm to 10 cm.

Write the word MONITOR the correct way up on the correct sides.

5 The box containing my pens is a cuboid with no lid.

The base is a square of side 6 cm and it is 15 cm high.

Draw the net for the box on squared paper. Use a scale of 1 cm to 3 cm.

Key ideas

● Accurate drawings must be accurate. Use a sharp pencil and check all measurements.

● The volume of a cuboid is length × width × height.

● When drawing a net make sure that edges that are to meet are the same length.

2 Rounding numbers and estimating

You should already know

- about place value, for example 378·46 is 3 hundreds, 7 tens, 8 units, 4 tenths and 6 hundredths
- how to multiply and divide using a calculator.

Approximating numbers

Rounding to the nearest 10, 100, ...

A number line may also help in rounding numbers to the nearest 10, 100 and so on.

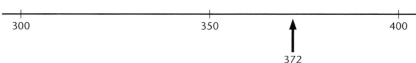

372 to the nearest 100 is 400.

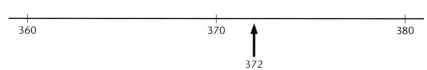

372 to the nearest 10 is 370.

Rounding to the nearest whole number or to a given number of decimal places (d.p.)

Think of a number line.

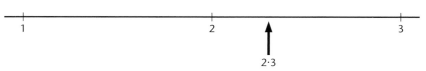

2·3 to the nearest whole number is 2.

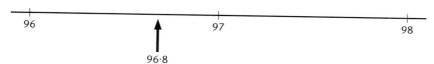

96·8 to the nearest whole number is 97.

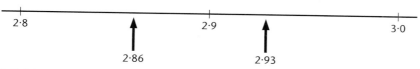

2·86 is nearer to 2·9 than 2·8 so is 2·9 to 1 d.p.
2·93 is nearer to 2·9 than 3·0 so is 2·9 to 1 d.p.

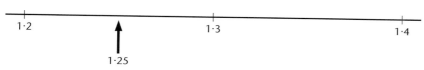

1·25 is midway between 1·2 and 1·3, but it is normal to count it up so it is 1·3 to 1 d.p.

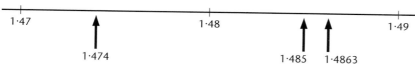

1·474 is nearer to 1·47 than 1·48 so is 1·47 to 2 d.p.
1·485 is midway between 1·48 and 1·49 so is 1·49 to 2 d.p.
1·4863 is nearer to 1·49 than 1·48 so is 1·49 to 2 d.p.

When rounding to a number of decimal places, count the digits after the decimal point.

You only change the final digit.

Exam tip

Remember the rule that if the value of the first digit being ignored is 5 or more then round up.
Examples:
3·75 is 3·8 to 1 d.p.
0·0136 is 0·014 to 3 d.p.

EXAMPLE 1

Write the following numbers correct to two decimal places (2 d.p.).

a) 9·368   **b)** 0·0438   **c)** 84·655.

a) 9·368

The first two decimal figures are 3 6, the third figure, the 8, is more than 5 so round up.

   = 9·37 (to 2 d.p.)

b) 0·0438

The first two decimal figures are 0 4, the third figure is 3, which is less than 5, so round down.

   = 0·04 (to 2 d.p.)

c) 84·655

The third decimal figure is 5 so the convention is to round up.

   = 84·66 (to 2 d.p.)

EXAMPLE 2

A cuboid has edges 95 cm, 75 cm and 130 cm.

Work out the volume of the cuboid. Give the answer to the nearest hundred.

Volume = 95 × 75 × 130

      = 926 250

      = 926 300 to the nearest 100

EXERCISE 2.1A

You can use a number line to help if you wish.

1 Write these numbers correct to the nearest whole number

   a) 2·7   **b)** 5·2
   c) 17·9   **d)** 231·4
   e) 19·5   **f)** 12·4
   g) 27·9   **h)** 99·6
   i) 16·5   **j)** 3815·2.

2 Write these numbers correct to the nearest 10.

   a) 456   **b)** 254
   c) 123   **d)** 998.

Write these numbers correct to the nearest 100.

   e) 5678   **f)** 9870
   g) 8801   **h)** 151.

Exercise 2.1A cont'd

3 Write these numbers correct to two decimal places.

a) 5·481   b) 12·0782
c) 0·214   d) 0·5666
e) 9·017   f) 78·044
g) 7·0064   h) 0·0734
i) 1·5236   j) 2·1256.

4 Write these numbers correct to 1 d.p.

a) 4·62   b) 5·47
c) 4·63   d) 8·41
e) 0·478   f) 0·1453
g) 82·16   h) 2·97
i) 6·15   j) 0·4512

5 Use your calculator to find the square root of 55. Write this value correct to one decimal place.

6 Work out these with a calculator, give the answers to three decimal places.

a) 1 ÷ 3   b) 2 ÷ 7
c) 3 ÷ 11   d) 4 ÷ 13
e) 28 ÷ 6

7 Add these numbers together and divide by 8.

Give your answer correct to one decimal place.

4, 6, 8, 9, 11, 3, 2, 15

8 A rectangle is 2·5 cm wide and 3·7 cm long. Find its area correct to 1 d.p.

9 A cube has a side of 4·82 cm.

Find its volume correct to 1 d.p.

10 (harder) Round each of the following to the stated accuracy.

a) 14·98 (1 d.p.)
b) 95·895 (2 d.p.)
c) 19·98 (1 d.p.)
d) 1019·996 (2 d.p.)
e) 49·95 (nearest whole number)
f) 199·95 (nearest 10).

EXERCISE 2.1B

1 Round these numbers to the nearest 10.

a) 97   b) 111
c) 374   d) 444

Round these numbers to the nearest 1000.

e) 1234   f) 8724
g) 6789   h) 9988

2 Write these numbers correct to the nearest whole number

a) 2·14   b) 10·98
c) 1·19   d) 2·31
e) 419·5   f) 12·92
g) 127·96   h) 199·4
i) 106·502   j) 52·514

Exercise 2.1B cont'd

3 Write these numbers correct to two decimals places.

   a) 9·424   **b)** 0·8413
   c) 0·283   **d)** 0·85
   e) 7·093   **f)** 18·63
   g) 7·1111   **h)** 8·081
   i) 4·656   **j)** 3·725.

4 Work out the following, give the answers to two decimal places.

   a) 1·38 × 6·77
   b) 4·7 × 3·65
   c) 9·125 ÷ 3·1
   d) 16·4 × 3·29.

5 Work out the following, give the answers to 1 d.p.

   a) 1 ÷ 8   **b)** 3 ÷ 7
   c) 4 ÷ 11   **d)** 5 ÷ 13.

6 Add the following together and divide by 7.

Give the answer to the nearest 10.

351, 521, 791, 831, 941, 1171, 1351.

7 A triangle has a base of 4·2 cm and a height of 6·7 cm. Find its area correct to 1 d.p.

8 A cuboid has sides 4·5 cm, 2·9 cm and 3·7 cm. Work out its volume correct to 2 d.p.

9 Six friends have a meal together and the bill comes to £28·71.

They divide the bill between them. What does it cost each of them correct to the nearest penny?

10 (harder) Round each of the following to the stated accuracy.

   a) 4·97 (1 d.p.)
   b) 19·915 (2 d.p.)
   c) 8·996 (2 d.p.)
   d) 49·996 (1 d.p.)
   e) 0·065 (1 d.p.)
   f) 65 995 (nearest 100).

Rounding to one significant figure (s.f.)

Often figures are rounded to 1 figure in newspapers.

'30 000 see Derby win' when the actual attendance was 29 877.

'Cambridge man wins £3 million' when the actual figure was £3 132 677.

To round to one significant figure (s.f.) all that is needed is to look at the second figure.

If it is less than 5 leave the first figure as it is.

If it is 5 or more add 1 to the first figure.

Then add zeros to show the size.

EXAMPLE 3

Write each of these to the numbers to one significant figure.

a) 5126   **b)** 5·817   **c)** 0·04715   **d)** 146 500   **e)** 1968.

Answers   **a)** 5000   Second figure is 1 so 5 stays as it is. Add three zeros to show the size.

        b) 6   Second figure is 8, so change 5 to 6. No zeros are needed.

        c) 0·05   Second figure is 7 so 4 becomes 5 and keep zero before the 5 to show the size.

        d) 100 000   Second figure is 4 so 1 stays as it is. Add zeros to show the size.

        e) 2000   Second figure is 9 so 1 changes to 2. Add zeros to show the size.

To estimate answers to problems, round each number to one significant figure and then carry out the calculation

EXAMPLE 4

Jim buys 24 golf balls at £3·75 each. What is the approximate cost?

Answer 20×4 (rounding to one significant figure)

       = £80.

Exam tip

Zeros are used to show size, a common error is to get the wrong number of zeros.

EXAMPLE 5

Estimate the value of these calculations by rounding to one significant figure.

a) 3·7 × 2·1   **b)** 47 × 63   **c)** 4·32 × 8·76   **d)** 2·95 × 0·321

e) 87 ÷ 38   **f)** 84·3 ÷ 6·72   **f)** 4·21 × 0·016   **g)** 0·815 ÷ 3·41

Answers   **a)** $4 \times 2 = 8$

        b) $50 \times 60 = 3000$

        c) $4 \times 9 = 36$

        d) $3 \times 0·3 = 0·9$

        e) $90 \div 40 = 2·25$ or 2·3 or 2

        f) $4 \times 0·02 = 0·08$

        g) $0·8 \div 3 = 0·2666 = 0·27$ or 0·3

$90 \div 40$ is the same as $9 \div 4 = 2·25$ answers should really only have one or two digits, so here round to one decimal place or one significant figure.

$2 \times 4 = 8$ and there are two decimal places to two decimal places.

EXERCISE 2.2A

1 Write each of these to 1 s.f.

   a) 4·2   **b)** 6·45
   c) 7·9   **d)** 4·25
   e) 5·6   **f)** 62·1
   g) 45·9   **h)** 26·5
   i) 314   **j)** 4579.

2 Write each of these to 1 s.f.

   a) 4271   **b)** 26·45
   c) 910·9   **d)** 64·25
   e) 0·002156 **f)** 6·21
   g) 4·59   **h)** 0·00265
   i) 0·00314   **j)** 0·0459.

For the rest of the questions show the approximations that you use to get the estimate.

3 Sally bought 18 packets of crisps at 32p each. Estimate how much she spent.

4 A cuboid has edges 3·65 cm, 2·44 cm and 2·2 cm. Estimate the volume.

5 Ian played 23 innings at cricket during the season. He scored a total of 734 runs. Estimate how many runs he scored on average per innings.

6 At Dave's ice cream stall he sold 223 ice creams at 67p each. Estimate how much he took.

7 Martyn rode 327 miles in 12 days. Estimate how far he rode on average each day.

8 Estimate the answers to these calculations.

   a) 5·89 × 1·86
   b) 19·25 ÷ 3·8
   c) 36·87 × 22·87
   d) 9·7 ÷ 3·5
   e) 2·14 × 0·82
   f) 3·14 × 7·92
   g) 289 ÷ 86
   h) 4·93 × 0·025.

EXERCISE 2.2B

1 Write each of these to 1 s.f.

   a) 7·21   **b)** 1·5
   c) 9·0   **d)** 5·4
   e) 1·56   **f)** 49·5
   g) 632   **h)** 7815
   i) 63   **j)** 972.

2 Write each of these to 1 s.f.

   a) 0·3721   **b)** 6·415
   c) 9034·9   **d)** 54·65
   e) 0·0156   **f)** 498·5
   g) 0·0567   **h)** 0·9815
   j) 0·1963   **f)** 0·97.

For the rest of the questions show the approximation that you use to get the estimate.

Exercise 2.2B cont'd

3 At the Arts Theatre seats cost £7·95 and 412 were sold for the production of 'Anything Goes'. Estimate the amount they took for the seats.

4 At a theatre they sold coffee in the interval at 68p a cup and sold 73 cups. Estimate the amount they took for coffee.

5 A rectangle has an area of 63·53 cm² and its length is 9·61 cm. Estimate the width of the rectangle.

6 On a long-haul flight an aeroplane flew 3123 miles in 8·4 hours. Estimate how far it flew in each hour on average.

7 To find the area of a circle of radius 2·68 metres. You need to work out $3·14 \times 2·68 \times 2·68$. Estimate the value of this calculation.

8 Estimate the answers to these calculations.

a) $14 \times 2·65$
b) 912×26
c) $54·6 \times 2·91$
d) $45 \div 6·8$
e) $287 \div 9·81$
f) $3·462 \times 0·34$
g) $0·27 \div 4·8$
h) $2·1 \times 9·4 \times 3·7$.

Key ideas

- 23 to the nearest 10 is 20.
- 379 to the nearest 100 is 400.
- 4·569 correct to two decimal places is 4·57.
- 0·0343 to 1 significant figure is 0·03.
- If the first figure being ignored is 5 or more then increase the last figure required by 1.
- To estimate the result of a calculation, round each of the figure to one significant figure and then carry out the calculation.

Metric and imperial units

You should already know

- that lengths are measured in kilometre (km), metre (m), centimetre (cm), millimetre (mm), inch, foot, yard and mile
- that masses are measured in tonne, kilogram (kg), gram (g), milligram (mg), ounce (oz) and pound (lb)
- that areas are measured in square units: mm², cm², m², km²
- that volumes are measured in cubic units, e.g. cm³, m³, litre (l), centilitre (cl), millilitre (ml), pint, gallon
- how to multiply and divide by powers of 10 with and without a calculator.

Converting between metric units

All these connections need to be learnt.

length   1 km = 1000 m, 1 m = 100 cm, 1 m = 1000 mm, 1 cm = 10 mm.

Mass   1 tonne = 1000 kg, 1 kg = 1000 g, 1 g = 1000 mg.

Capacity   1 litre = 100 cl, 1 litre = 1000 ml. 1 ml is the same as 1 cm³.

Area $1 \text{ m}^2 = 100 \text{ cm} \times 100 \text{ cm} = 10\,000 \text{ cm}^2$, $1 \text{ cm}^2 = 10 \text{ mm} \times 10 \text{ mm} = 100 \text{ mm}^2$.

Exam tip

A help to learn these connections is that kilo means 1000, centi means $\frac{1}{100}$, milli means $\frac{1}{1000}$.

To change from one unit to another you need to multiply or divide by the connection. For example to change from metres to centimetres you need to multiply by 100. To change from centimetres to metres you need to divide by 100.

EXAMPLE 1

Change the following.
a) 14 m to cm   **b)** 25 cm to mm
c) 2·8 km to m   **d)** 2·85 kg to g
e) 12·4 cm² to mm²   **f)** 5·9 litre to cm³.

a) 14 m = 14 × 100 cm = 1400 cm
b) 25 cm = 25 × 10 mm = 250 mm
c) 2·8 km = 2·8 × 1000 m = 2800 m
d) 2·85 kg = 2·85 × 1000 g = 2850 g
e) 12·4 cm² = 12·4 × 100 mm² = 1240 mm²
f) 5·9 litre = 5·9 × 1000 cm³ = 5900 cm³ (or ml).

EXAMPLE 2

Change the following.
a) 2500 cm to m   **b)** 48 mm to cm
c) 85 000 m to km   **d)** 2400 g to kg
e) 940 ml to litres **f)** 48 000 cm² to m²

a) 2500 cm = 2500 ÷ 100 m = 25 m
b) 48 mm = 48 ÷ 10 cm = 4·8 cm
c) 85 000 m = 85 000 ÷ 1000 km = 85 km
d) 2400 g = 2400 ÷ 1000 kg = 2·4 kg
e) 940 ml = 940 ÷ 1000 litres = 0·94 litres.
f) 48 000 cm² = 48 000 ÷ 10 000 m² = 4·8 m².

Exam tip

A whole number has a decimal point after the last digit.

EXERCISE 3.1A

Change the following.

1 **a)** 25 m to cm
   b) 42 cm to mm
   c) 2·36 m to cm
   d) 5·1 m to mm

2 **a)** 900 cm to m
   b) 4600 mm to cm
   c) 931 mm to cm
   d) 84 600 cm to m

3 **a)** 5 kg to g
   b) 4·16 tonne to kg
   c) 2 g to mg
   d) 5 litres to ml

4 **a)** 2000 g to kg
   b) 9200 ml to litre
   c) 5740 kg to tonne
   d) 764 g to kg

5 **a)** 3 m² to cm²
   b) 2·3 cm² to mm²
   c) 9·42 m² to cm²
   d) 0·014 cm² to mm²

6 **a)** 90 000 mm² to cm²
   b) 8140 mm² to cm²
   c) 7 200 000 cm² to m²
   d) 94 000 cm² to m²

Exercise 3.1A cont'd

7 a) 640 cm to mm
   b) 2100 mm to m
   c) 42·3 mm to cm
   d) 524 cm to m

8 a) 9·4 litres to ml
   b) 154·2 g to kg
   c) 9·1 kg to g
   d) 5·4 kg to tonne

9 a) $1 \cdot 5 \, m^2$ to mm^2
   b) $916\,400 \, cm^2$ to m^2
   c) $8100 \, mm^2$ to cm^2
   d) $0 \cdot 86 \, m^2$ to cm^2

10 a) $2 \cdot 61$ litre to cm^3
   b) 9500 ml to litre
   c) 2·4 litre to ml
   d) 910 ml to litre

11

a) This girl is 1·04 metres tall. How tall is she in centimetres?
b) This book has a mass of 1214 grams. How much is that in kilograms?

12 This bottle holds 2 litres of lemonade. To fill a small glass needs 80 ml. How many of these glasses could be filled from this bottle?

EXERCISE 3.1B

Change the following

1 a) 52 m to cm
   b) 24 cm to mm
   c) 12·6 m to cm
   d) 0·15 m to mm

2 a) 1100 cm to m
   b) 2500 mm to cm
   c) 314 mm to cm
   d) 84·6 cm to m

Exercise 3.1B cont'd

3
  a) 1.2 kg to g
  b) 3.6 tonne to kg
  c) 9 g to mg
  d) 1.5 litres to cm^3

4
  a) 9000 g to kg
  b) 7200 ml to litre
  c) 140 kg to tonne
  d) 640 g to kg

5
  a) 12 m^2 to cm^2
  b) 3.71 cm^2 to mm^2
  c) 0.42 m^2 to cm^2
  d) 0.05 cm^2 to mm^2

6
  a) $120\,000$ mm^2 to cm^2
  b) $38\,400$ mm^2 to cm^2
  c) 4100 cm^2 to m^2
  d) $421\,000$ cm^2 to m^2

7
  a) 20 cm to mm
  b) $13\,400$ mm to m
  c) 2.35 m to cm
  d) 2.42 cm to m

8
  a) 4 litres to ml
  b) 5420 g to kg
  c) 0.21 kg to g
  d) 5400 kg to tonne

9
  a) 3 m^2 to mm^2
  b) $412\,500$ cm^2 to m^2
  c) 9400 mm^2 to cm^2
  d) 0.06 m^2 to cm^2

10
  a) 2.13 litres to cm^3
  b) 5100 ml to litre
  c) 421 litre to ml
  d) 91.7 ml to litre

11

The bucket contains 1346 g of sand. How much more can be added without overloading it?

12 A shelf is 1.8 m long. How many books 25 mm wide will fit on the shelf?

Metric and imperial units

You need to learn these approximate equivalents between metric and imperial units.

Length	8 km = 5 miles	1 m = 40 inches	30 cm = 1 foot.
Mass	1 kg = 2·2 lb.		
Capacity	1 litre = 1·75 pint, 4·5 litres = 1 gallon.		

EXAMPLE 3

Stacey bought a 5 kg bag of sand. What is its approximate mass in pounds?

5 kg = 5 × 2·2 lb = 11 lb.

EXAMPLE 4

The distance from Bristol to Cardiff is 44 miles.
Approximately how far is it in kilometres?

44 miles = $44 \times \frac{8}{5}$ = 70.4 = 70 km.

8 km = 5 miles so to change from miles to km multiply by 8 and then divide by 5.

EXAMPLE 5

My bucket holds 5 gallons of water. What is this in litres?

5 gallon = 5×4.5 litres = 22.5 (or 23 litres).

Exam tip

If you are not sure wich way to multiply or divide remember which is the shorter unit of distance, mass or capacity and that you will have more of that one.

Do not try to be too accurate with an approximate answer. In Example 4 70 km was accurate enough.

EXERCISE 3.2A

In all these questions use the approximate equivalents and give the answers to a sensible degree of accuracy.

1

Tony walked 15 miles on a hike. How far was this in kilometres?

2 Stephen's mass is 62 kg. What is his mass in pounds?

3 Jean bought 8 litres of drink for her party. How many pints was this?

4 The height of my desk is 1·2 metres. How high is this in inches?

5 In 1960 the long jump record was about 26 feet.

   a) How far was this in centimetres?
   b) Change your answer to metres.

Exercise 3.2A cont'd

6 The Scruton household drink 24 pints of milk a week. How many litres is this?

7

Mrs Benson bought 8 lb of potatoes, 2 lb of carrots, 1 lb of peas and 3 lb of onions. What was the total mass of these vegetables altogether in kilograms?

8 Keri drove 128 km along the motorway. How far was this in miles?

9

To paint the outside of her house Doreen was told she needed 15 litres of paint. How many pints is this?

10 A cross-country race was 15 miles long. How far was it in kilometres?

EXERCISE 3.2B

In all these questions use the approximate equivalents and give the answers to a sensible degree of accuracy.

1 The distance from Lincoln to Liverpool is 224 km. How far is it in miles?

2 Margaret bought a 5 litre container of disinfectant. How many pints was it?

3

The total mass of the luggage of 25 people on a trip to Barcelona was 500 kg. What was the mass in pounds?

Chapter 3 *Metric and imperial units*

Exercise 3.2B cont'd

4 Holly washed her car with six buckets full of water. Each bucket contained 20 litres of water. How many gallons of water did she use altogether?

5

This car can travel 370 miles on a tank full of petrol. The distance from Manchester to Oxford is 256 km.

a) Could you drive to Oxford and back on one tank of petrol?

b) How many miles are you short or how many miles more could you drive?

6 Potatoes can be bought in 56 lb bags. How many kilograms is this?

7 To fill his pond, Karl used 14 gallons of water. How many litres was this?

8 The height of David's house is 18 feet.

a) What is the height in centimetres?

b) What is the height in metres?

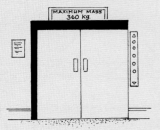

9 The maximum mass that a lift can carry is 360 kg. How many pounds is this?

10 Darren was told that he was 1·74 m tall.

a) How many inches is this?

b) There are 12 inches in a foot. Change the answer to feet and inches.

Key ideas

- 1 km = 1000 m, 1 m = 100 cm, 1 m = 1000 mm, 1 cm = 10 mm.
- 1 tonne = 1000 kg, 1 kg = 1000 g, 1 g = 1000 mg.
- 1 litre = 100 cl, 1 litre = 1000 ml. 1 ml is the same as 1 cm^3.
- 1 m^2 = 10 000 cm^2, 1 cm^2 = 100 mm^2.
- 8 km = 5 miles, 1 m = 40 inches, 30 cm = 1 foot.
- 1 kg = 2·2 lb.
- 1 litre = 1·75 pint, 4·5 litres = 1 gallon.
- To change between any of the units, multiply or divide by the connection.

Chapter 3   *Metric and imperial units*

4 Solving equations

Two-step equations

The expression $2x - 3$ means multiply x by 2 then subtract 3.
Thus the equation $2x - 3 = 5$ can be written as a flow chart

$$x \xrightarrow{} \boxed{\times 2} \xrightarrow{2x} \boxed{-3} \xrightarrow{} 2x - 3 = 5$$

To solve the equation all that is needed is to reverse the flow chart.

$$\xleftarrow{4} \boxed{\div 2} \xleftarrow{8} \boxed{+3} \xleftarrow{} 5$$

As you can see the order is reversed and so is every operation. This gives $x = 4$.

EXAMPLE 1

Use a flow diagram to solve these equations.

a) $3x + 2 = 11$ **b)** $5x - 4 = 8$

a) $$x \xrightarrow{} \boxed{\times 3} \xrightarrow{3x} \boxed{+2} \xrightarrow{} 3x + 2 = 11$$

$$\xleftarrow{3} \boxed{\div 3} \xleftarrow{9} \boxed{-2} \xleftarrow{} 11$$

so $x = 3$.

b) $$x \xrightarrow{} \boxed{\times 5} \xrightarrow{5x} \boxed{-4} \xrightarrow{} 5x - 4 = 8$$

$$\xleftarrow{\frac{12}{5}} \boxed{\div 5} \xleftarrow{12} \boxed{+4} \xleftarrow{} 8$$

so $x = \frac{12}{5}$ or $2\frac{2}{5}$. Either answers is acceptable and to get the second you divide 5 into 12 and 2 is left over.

EXERCISE 4.1A

Use flow charts to solve these equations.

1	$3x + 1 = 7$	**6**	$6x - 1 = 5$
2	$5x + 2 = 7$	**7**	$5x - 2 = 8$
3	$3x - 2 = 7$	**8**	$3x + 5 = 17$
4	$2x + 4 = 12$	**9**	$5x - 4 = 11$
5	$4x + 1 = 13$	**10**	$2x - 1 = 6$

EXERCISE 4.1B

Use flow charts to solve these equations.

1	$2x - 1 = 7$	**6**	$5x - 11 = 24$
2	$5x + 1 = 11$	**7**	$6x - 2 = 10$
3	$3x + 2 = 11$	**8**	$2x + 5 = 10$
4	$2x + 7 = 9$	**9**	$5x + 2 = 11$
5	$3x + 1 = 13$	**10**	$12x - 5 = 31$

Another way to solve these equations is shown in this example.

EXAMPLE 2

Solve the equation $3x - 1 = 5$.

This needs two steps, doing the opposite operation each time.

$$3x - 1 = 5$$
$[3x - 1 + 1 = 5 + 1]$ Add 1 to each side.
$$3x = 6$$
$[3x \div 3 = 6 \div 3]$ Divide each side by 3.
$$x = 2.$$

Exam tip

Remember, all the operations must be done to all of each side.

EXERCISE 4.2A

Solve these equations.

1	$2x + 1 = 3$	**6**	$5x + 2 = 7$
2	$2x - 1 = 3$	**7**	$7x - 3 = 18$
3	$2x + 3 = 1$	**8**	$4x + 7 = 3$
4	$2x - 3 = 1$	**9**	$2x - 3 = 4$
5	$3x + 2 = 8$	**10**	$3x + 2 = 9$

EXERCISE 4.2B

Solve these equations.

1	$2x + 5 = 9$	**6**	$11x - 4 = 7$
2	$2x - 5 = 9$	**7**	$10x + 3 = 18$
3	$3x + 7 = 4$	**8**	$2x - 7 = 4$
4	$3x - 7 = 8$	**9**	$2x + 7 = 4$
5	$4x - 11 = 5$	**10**	$5x - 3 = {}^-5$

In all the equations you have just done the unknown letter was on the left-hand side of the equation. It can appear on either side and in later work on both sides.

If it just appears on the right-hand side, it is easiest to change the sides round and work as above.

EXAMPLE 3

Solve the equation $14 = 5x + 2$.

$5x + 2 = 14$	Exchange sides.
$[5x + 2 - 2 = 14 - 2]$	Subtract 2 from each side.
$5x = 12$	
$[5x \div 5 = 12 \div 5]$	Divide both sides by 5.
$x = 2\frac{2}{5}$ or 2·4.	

EXAMPLE 4

Solve the equation $3 = 7 + 2x$

$7 + 2x = 3$	Exchange sides.
$[7 + 2x - 7 = 3 - 7]$	Subtract 7 from both sides.
$2x = {}^-4$	
$[2x \div 2 = {}^-4 \div 2]$	Divide both sides by 2.
$x = {}^-2.$	

Exam tip

A common wrong answer for $5x = 12$ is $x = \frac{5}{12}$ rather than $\frac{12}{5}$.

EXERCISE 4.3A

Solve the equations.

1. $13 = 3x + 1$
2. $15 = 2x - 3$
3. $8 = 4x - 2$
4. $5 = 15 + 2x$
5. $10 = 3x - 5$
6. $20 = 2x + 6$
7. $7 = 6x + 25$
8. $4 = 3x - 5$
9. $7x - 2 = 12$
10. $15 = 4x + 5$

EXERCISE 4.3B

Solve these equations.

1. $3 = 2x + 1$
2. $25 = 4x - 3$
3. $5 = 3x - 1$
4. $1 = 10 + 3x$
5. ${}^-7 = 2x + 5$
6. $15 = 2x + 6$
7. $13 = 4x - 7$
8. $8 = 3x + 11$
9. $6x - 4 = 14$
10. $20 = 5x + 2$

You may have equations in which the term in x is negative. These will involve an extra step.

EXAMPLE 5

Solve the equation $13 - 5x = 3$

$[13 - 5x + 5x = 3 + 5x]$	Add $5x$ to both sides.
$13 = 3 + 5x$	
$3 + 5x = 13$	Exchange sides.
$[3 + 5x - 3 = 13 - 3]$	Subtract 3 from both sides.
$5x = 10$	
$[5x \div 5 = 10 \div 5]$	Divide both sides by 5.
$x = 2.$	

These exercises contains a mixture of all the equations you have covered in this chapter.

EXERCISE 4.4A

Solve these equations.

1 $13 - 3x = 1$ **6** $2 = 2x + 6$

2 $1 = 3 - 2x$ **7** $7 = 19 - 6x$

3 $8 = 4x + 12$ **8** $1 = 3x - 5$

4 $2x - 7 = 9$ **9** $12 - 7x = 12$

5 $10 = 18 - 3x$ **10** $4x + 15 = 7$

EXERCISE 4.4B

Solve these equations

1 $11 - 2x = 7$ **6** $4 = 3x - 6$

2 $3 = 13 - 2x$ **7** $4 = 19 + 5x$

3 $8 = 2x + 7$ **8** $1 = 3x + 5$

4 $5x - 3 = 9$ **9** $2 - 7x = 9$

5 $4 = 13 - 3x$ **10** $3x + 9 = 2$

Key idea

● **The two sides of an equation must be kept equal. Operations carried out to simplify or solve an equation must always be the same for each side.**

Revision exercise

1 a) Find the volume of these boxes, all lengths are in centimetres.

(i)

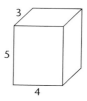

(ii)

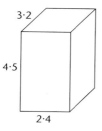

b) Draw a net for each of the boxes. They both have lids.

2 Write the following numbers correct to two decimal places.

a) 7·897 **b)** 13·1234
c) 0·243 **d)** 0·6772.

3 Write these numbers to the nearest 10.

a) 127 **b)** 543
c) 995 **d)** 1239.

Write these numbers to the nearest 100.

e) 7898 **f)** 9820
g) 8850 **h)** 51.

4 A cuboid has dimensions 4·3 by 5·2 by 6·7.

a) Find the volume, write down all the figures on your calculator

b) Write the answer correct to one decimal place.

5 Write these numbers correct to one significant figure.

a) 3·27 **b)** 145
c) 9471 **d)** 1·5
e) 65·7 **f)** 14·5
g) 584·2 **h)** 0·52
i) 0·028 **j)** 791 000

6 Do not use a calculator in this question. Show your working.

a) In the 28 days of February Beth earned £864. Estimate how much she earned each day.

b) At the motorway service station the bottles of soft drink were priced at 73p. In one day they sold 578 bottles. Estimate how much they took for the soft drinks.

7 Write these amounts in the units indicated.

a) 3·2 m in mm
b) 4·5 m in cm
c) 14 523 mm in m
d) 584·2 cm in m
e) 9471 g in kg
f) 20 litre in ml
g) 500 cl in litre
h) 4800 ml in litre.

8 In each part of this question, put the measures in order of size.

  a) 5 m, 500 mm, 655 cm, 7124 mm, 2·375 m

  b) 15 m, 2·4 km, 1500 m, 5000 m, 1·7 km

  c) 4000 g, 5·2 kg, 1·004 kg, 5020 g, 5·002 kg

  d) 5 *l*, 497 cl, 5010 ml, 2·943 *l*, 96·42 cl.

9 Change these amounts in imperial units to their rough metric equivalents.

  a) 5 inches   **b)** 3 feet
  c) 10 miles   **d)** 56 lb
  e) 6 pints   **f)** 3 gallons.

10 Change these amounts in metric units to their rough imperial equivalents.

  a) 10 cm   **b)** 1·5 m
  c) 36 km   **d)** 3 kg
  e) 1·5 litres **f)** 20 litres.

11 Solve these equations.

  a) $4x + 7 = 11$
  b) $19 = 13 + 2x$
  c) $4 = 3x - 2$
  d) $24 = 19 + 5x$
  e) $5 = 2x + 1$
  f) $5x + 9 = 4$
  g) $11 = 3x + 5$
  h) $12 + 4x = 0$

Powers, roots, multiples and factors

You should already know

the meaning of each of these words:
- digit
- factor
- multiple
- consecutive
- square number
- square root
- cube
- cube root.

Powers and roots

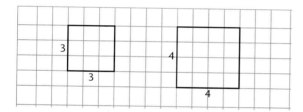

As you can see the square with side 3 has area $3 \times 3 = 9$, and the square with side 4 has area $4 \times 4 = 16$. These can be stated as the square of 3 is 9 and written as $3^2 = 9$ etc.

The set of whole numbers, 1, 4, 9, 16, 25, 36, ... are **squares** of the counting numbers.

Because $16 = 4^2$ the **square root** of 16 is 4, written as $\sqrt{16} = 4$. Similarly $\sqrt{36} = 6$ and $\sqrt{81} = 9$.

The numbers 1, 8, 27, 64, ... are **cube** numbers because each of them can be written as the cube of a whole number, for example $8 = 2^3$, $27 = 3^3$.

Because $27 = 3^3$ the **cube root** of 27 is 3, written as $\sqrt[3]{27} = 3$.

You can use your calculator to find squares, square roots, cubes and cube roots. Learn which buttons you need to use to do these operations. You have already seen that 5^2 means 5×5, 5^3 means $5 \times 5 \times 5$. In a similar way 5^4 means $5 \times 5 \times 5 \times 5$.

EXERCISE 5.1A

1 Write down the square of each number.

a) 7   b) 12   c) 25
d) 40   e) 9.

2 Write down the square root of each number.

a) 49   b) 121   c) 169
d) 289   e) 625.

3 Write down the cube of each number.

a) 4   b) 5   c) 6
d) 10   e) 12.

4 Write down the cube root of each number.

a) 27   b) 343   c) 729
d) 1331   e) 1 000 000.

5 Work out

a) 5^2 b) 7^2 c) 11^2
d) 16^2 e) 18^2.

6 Work out
a) $\sqrt{81}$ b) $\sqrt{121}$ c) $\sqrt{225}$
d) $\sqrt{625}$ e) $\sqrt{169}$.

7 Work these out, giving your answers correct to two decimal places.
a) $\sqrt{56}$ b) $\sqrt{27}$ c) $\sqrt{60}$
d) $\sqrt{280}$ e) $\sqrt{678}$.

8 Work out
a) 4^2 b) 2^3 c) 7^2
d) 5^2 e) 10^4.

9 Work out
a) $6^2 - 5^2$ b) $2^2 + 3^2$
c) $8^2 - 4^2$ d) $3^2 - 2^3$
e) $5^2 - 4^2 - 3^2$.

10 Work out $\sqrt{12^2 - 8^2}$ give the answer to 3 s.f.

EXERCISE 5.1B

1 Write down the square of each number.

a) 8   b) 11   c) 35
d) 50   e) 21.

2 Write down the square root of each number.

a) 81   b) 144   c) 196
d) 361   e) 576.

3 Write down the cube of each number.

a) 2   b) 3   c) 1·5
d) 7   e) 2^2.

4 Write down the cube root of each number.

a) 125   b) 216   c) 512
d) 1728   e) 1000.

5 Work out

a) 6^2 b) 9^2 c) 13^2
d) 22^2 e) 32^2.

Exercise 5.1B cont'd

6 Work out

  a) $\sqrt{100}$ **b)** $\sqrt{256}$ **c)** $\sqrt{324}$

  d) $\sqrt{841}$ **e)** $\sqrt{784}$.

7 Work these out, giving your answers correct to two decimal places.

  a) $\sqrt{70}$ **b)** $\sqrt{39}$ **c)** $\sqrt{90}$

  d) $\sqrt{380}$ **e)** $\sqrt{456}$.

8 Calculate

  a) 5^2 **b)** 4^3 **c)** 8^2

  d) 10^3 **e)** 10^7.

9 Work out

  a) $4^2 + 5^2$ **b)** $6^2 - 3^2$

  c) $21^2 - 9^2$ **d)** $25^2 - 5^3$

  e) $13^2 - 12^2 - 5^2$

10 Work out $\sqrt{17^2 - 16^2}$, give the answer to 3 s.f.

You can work out all the powers and roots using your calculator, but it is useful to know at least the basic squares. These are $1^2 = 1$, $2^2 = 4$, $3^2 = 9$, $4^2 = 16$, $5^2 = 25$, $6^2 = 36$, $7^2 = 49$, $8^2 = 64$, $9^2 = 81$, $10^2 = 100$.

The rest of this chapter is a series of investigations which use the knowledge needed about multiples, factors and roots.

ACTIVITY 1

Consecutive numbers

This investigation into consecutive numbers is given in two versions. Version A is set as a structured task, version B is an unstructured task. The solutions and comments for both versions are also given below. Work through the solution and comments. Ask your teacher if you are not sure of anything.

Version A

It is claimed that it is possible to make any number by adding consecutive numbers.

For example:

$2 + 3 = 5$

$2 + 3 + 4 = 9$

$2 + 3 + 4 + 5 = 14$.

1 Can all odd numbers be written as the sum of two consecutive numbers?

2 Can even numbers be written as the sum of two consecutive numbers?

3 Investigate any relationship between the number of consecutive numbers added together, and their sum.

4 Investigate any other patterns you notice.

Version B

$15 = 7 + 8 = 4 + 5 + 6$

$\quad = 1 + 2 + 3 + 4 + 5$.

Investigate consecutive numbers.

Solution and comments for Version A

Question 1

1

$3 = 1 + 2$

$5 = 2 + 3$

$7 = 3 + 4$

$9 = 4 + 5$

$11 = 5 + 6$

This suggests that all odd numbers can be written as the sum of two consecutive numbers.

Test: $63 = 31 + 32$

Proof: If the first number is n then the consecutive number is $n + 1$. The sum is $n + (n + 1) = 2n + 1$. Doubling a number gives an even number, adding 1 then will always make the answer odd. Therefore adding two consecutive numbers will always give an odd answer.

Question 2

Even numbers are 2, 4, 6, 8, …

Adding two consecutive numbers always gives an odd number (see the notes above), so it is not possible to make an even number in this way.

Proof: As stated in question 1 above, $n + (n + 1)$ will always be odd.

In the answers to these two questions you are showing that you can:

- organise work, check results
- find out necessary information
- discuss work
- show you understand the task
- find examples which fit a statement
- search for a pattern, using at least three results
- explain your reasoning and make statements about the results you found.

Exam tip

Start with the smallest odd number and try to match each odd number to an addition sum. Work systematically, write the results in order.

Question 3

Experiment with three consecutive numbers:

$1 + 2 + 3 = 6$

$2 + 3 + 4 = 9$

$3 + 4 + 5 = 12$

With four consecutive numbers:

$1 + 2 + 3 + 4 = 10$

$2 + 3 + 4 + 5 = 14$

$3 + 4 + 5 + 6 = 18$

So with five consecutive numbers will the total increase by 5 each time?

Check:

$1 + 2 + 3 + 4 + 5 = 15$

$2 + 3 + 4 + 5 + 6 = 20$

$3 + 4 + 5 + 6 + 7 = 25$

The prediction is correct and the numbers are in the 5 times table.

Work systematically, write the results clearly. These results suggest that the numbers go up in 3s, and are in the 3 times table.

These results suggest the numbers go up in 4s.

Think!

… the totals might go up in steps equal to the number of numbers added together.

The skills shown are the same as in questions 1 and 2, together with:

● testing a generalisation.

The question said, 'Investigate any relationship between the number of consecutive numbers added together, and their sum.' Has this been answered? (Yes – because any other patterns seen can be followed up in question 4.)

Question 4

a) With three consecutive numbers, the sum is 3 times the middle number.

$3 + 4 + 5 = 12 = 3 \times 4$

$5 + 6 + 7 = 18 = 3 \times 6$

Proof: If the three numbers are n, $n + 1$, $n + 2$ then:

$$n + n + 1 + n + 2 = 3n + 3$$
$$= 3(n + 1)$$

i.e. three times the middle number. Thus the numbers are in the 3 times table, 3 is a factor of the totals.

● explaining your reasoning and making a statement about the results found

● beginning to justify solutions, explaining why the results occur

To find three consecutive numbers that make a given total, divide the total by 3, then add 1 and subtract 1 to find all three numbers.

Example

Which three consecutive numbers make 72?

$72 \div 3 = 24$, so the numbers are

$24 - 1 = 23$, 24 and $24 + 1 = 25$.

b) For four consecutive numbers, the sum is twice the sum of the middle two numbers.

$1 + 2 + 3 + 4 = 10 = 2 \times (2 + 3)$

$3 + 4 + 5 + 6 = 18 = 2 \times (4 + 5)$

Explaining your reasoning and making a statement about the results found

Proof:

$n + n + 1 + n + 2 + n + 3$

$\quad = n + (n + 1 + n + 2) + n + 3$

$\quad = n + (2n + 3) + n + 3$

$\quad = (2n + 3) + (2n + 3)$

$\quad = 2(2n + 3)$

Beginning to justify solutions, explaining why the results occur

c) For five consecutive numbers the totals will be in the the 5 times table and will be 5 times the middle number.

d) The following numbers cannot be given as the sum of consecutive numbers: 2, 4, 8, 16, …

Each number is double the previous number – they are powers of 2:

$4 = 2 \times 2 = 2^2$

$8 = 2 \times 2 \times 2 = 2^3$

$16 = 2 \times 2 \times 2 \times 2 = 2^4$

The next number will be 3^2 which is 2^5.

Can you think of a reason for this?

Hint: Take n as the middle number, then write down the numbers on either side of n (i.e. $n - 1$ and $n + 1$).

Explaining your reasoning and making a statement about the results found

Can you think of a reason why this happens?

Think about the factors of, for example

16: 1, 2, 4, 8, 16

and of 20: 1, 2, 4, 5, 10, 20

What do you notice?

Comments for Version B

Because this task is set in an unstructured form you will need to think carefully and systematically about how you will work through it and show your results.

There are two possible approaches:

1 start with the totals, i.e. 1, 2, 3, 4, … and try to make them with sums of consecutive numbers
2 start with the actual sums, i.e. 1 + 2, 1 + 2 + 3, … and see what totals you can make.

Approach 1 is probably easier, and you are more likely to be systematic in your working than with approach 2. Approach 1 also allows you to see patterns more clearly, especially if you show the results in a table.

This table, for the numbers up to 30, is not complete.

Copy it and complete it to show that you can see all the patterns.

1			
2			
3	1 + 2		
4			
5	2 + 3		
6		1 + 2 + 3	
7	3 + 4		
8			
9	4 + 5	2 + 3 + 4	
10			1 + 2 + 3 + 4
11	5 + 6		
12		3 + 4 + 5	
13	6 + 7		
14			2 + 3 + 4 + 5
15	7 + 8	4 + 5 + 6	1 + 2 + 3 + 4 + 5
16			
17	8 + 9		
18		5 + 6 + 7	3 + 4 + 5 + 6
19			
20			2 + 3 + 4 + 5 + 6
21			
22		4 + 5 + 6 + 7	
23			
24			
25			3 + 4 + 5 + 6 + 7
26			
27			
28			
29			
30			

However a task is presented it is a good idea to try to extend it by asking some questions of your own.

Here are three possible extension questions to ask if you are attempting version B. Work through them and/or any ideas of questions of your own.

ACTIVITY 2

Adding consecutive numbers

a) What happens if you add two consecutive odd numbers i.e. 3 + 5, 5 + 7?

b) What happens if you add three consecutive odd numbers?

c) What happens if you add four consecutive odd numbers?

Look back at the table on the previous page. What do you notice about those numbers that can only be made by adding two consecutive numbers?

ACTIVITY 3

Differences between squares

What do you notice if

a) you write down two consecutive numbers, square both these numbers and find the difference between the squares

b) you write down three consecutive numbers, square the middle one, multiply the first and the third numbers together and find the differences between the answers?

Can you prove either or both of the results?

ACTIVITY 4

An investigation into squares

a) How many squares are there in each diagram?

b) How many squares will be there in the sixth diagram? and how many in the tenth?

      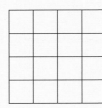

     A   **B**   **C**

ACTIVITY 5

3s and 5s

Some multiples of 3 are:

$9 = 3 \times 3$ or $3(3)$
$12 = 4 \times 3$ or $4(3)$
$30 = 10(3)$.

Similarly some multiples of 5 are:

$25 = 5(5)$
$40 = 8(5)$
$60 = 12(5)$.

Some numbers are multiples both of 3 and 5:

$30 = 10(3)$ and $6(5)$.

Some numbers can be made by adding a multiple of 3 to a multiple of 5:

$17 = 12 + 5 = 4(3) + 1(5)$

but some other numbers, such as 4, cannot be made at all. Investigate 3s and 5s.

ACTIVITY 6

The jail problem

- A high security jail has 100 prisoners.
- There is only one prisoner in each cell.
- The cells are numbered 1, 2, 3, …, 99, 100.
- Each cell has a guard.
- The guards share the duties but one night, as they go off duty, they make a mistake:

Guard 1 goes along the cells and unlocks everyone.

Guard 2 then locks the cells which are multiples of 2 i.e. cells 2, 4, 6, …

Guard 3 then visits all the cells that are multiples of 3, i.e. 3, 6, 9, … and unlocks any doors that are locked, and locks any door that is unlocked.

All the guards repeat this process as they go off duty, one by one.

After the last guard has left, which prisoners can escape? Explain why.

ACTIVITY 7

Number patterns

Look at this number pattern:

$49 \rightarrow 9^2 - 4^2 = 81 - 16 = 65$

$65 \rightarrow 6^2 - 5^2 = 36 - 25 = 11$

$11 \rightarrow 1^2 - 1^2 = 1 - 1 = 0$

Investigate two-digit numbers. What happens if you add the squares of the digits rather than finding the difference between them?

ACTIVITY 8

Gnomons

Here is part of a multiplication square. The reversed L-shape is known as a gnomon. The numbers in each gnomon in the table add up to a cube number. For example, 4^3 is equal to

$4 + 8 + 12 + 16 + 12 + 8 + 4 = 64$

$1^3 \rightarrow$	1	2	3	4	5
$2^3 \rightarrow$	2	4	6	8	10
$3^3 \rightarrow$	3	6	9	12	15
$4^3 \rightarrow$	4	8	12	16	20
$5^3 \rightarrow$	5	10	15	20	25

a) Now work out the values for 3^3 and 5^3.

Note:

$4 + 8 + 12 + 16 + 12 + 8 + 4$ can be written as

$4 \times 1 + 4 \times 2 + 4 \times 3 + 4 \times 4 + 4 \times 3 + 4 \times 2 + 4 \times 1$ and as 4 is clearly a common factor, this expression could be written as $4(1 + 2 + 3 + 4 + 3 + 2 + 1)$.

b) Try to write down the equivalent expressions for 3^3 and 5^3.

c) Check that this method works for 10^3.

Also note that:

$1^3 = 1^2 = 1$

$1^3 + 2^3 = (1 + 2)^2 = 9$

$1^3 + 2^3 + 3^3 = (1 + 2 + 3)^2 = 36.$

d) Work out as simply as possible

$1^3 + 2^3 + 3^3 + 4^3 + \ldots + 10^3.$

Finally, some more work on roots:

Whole numbers such as 1, 4, 9, 16, 25, … are square numbers, because, for example, $16 = 4 \times 4$ or '4 squared'.

The square root of 16, written as $\sqrt{16} = 4$. Similarly $\sqrt{25} = 5$ and $\sqrt{9} = 3$.

You should know the square roots of the square numbers from 1 to 100.

The numbers 1, 8, 27, 64, … are called cube numbers, because they can be written as the cube of a whole number, for example $27 = 3 \times 3 \times 3$ or '3 cubed', written as 3^3.

Because $64 = 4 \times 4 \times 4$ or 4^3 the cube root of 64, $\sqrt[3]{64} = 4$ and the cube root of 8, $\sqrt[3]{8} = 2$.

ACTIVITY 9

Square and cube roots

Write down the values of:

1   a) $\sqrt{81}$ b) $\sqrt[3]{125}$
    c) $\sqrt{289}$ d) $\sqrt[3]{216}$.

2   a) $\sqrt{(3^2 + 4^2)}$ b) $\sqrt{(169 - 144)}$
    c) $\sqrt{(5^2 + 12^2)}$.

Key ideas

- Factors are numbers that will divide into other numbers, so 3 is a factor of 9, but not a factor of 8.

- A multiple of a number will be in that number's 'times table', so 9 is a multiple of 3, 35 is a multiple of 5 and also 7.

- 2^3 means $2 \times 2 \times 2 = 8$

- 10^5 means $10 \times 10 \times 10 \times 10 \times 10 = 100\,000$.

6 Maps, bearings and scale drawings

You should already know

- how to use a ruler and protractor
- the basic co-ordinates
- the basic compass points.

Grid references

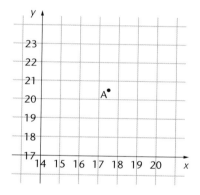

The point A is in the square where the bottom left-hand corner has co-ordinates (17, 20).

The point A is said to have four-figure grid reference 17 20.

You need to follow directions on a map and give basic compass directions.

EXAMPLE 1

This is a sketch map of an island.

Scale = 2 cm to 1 mile

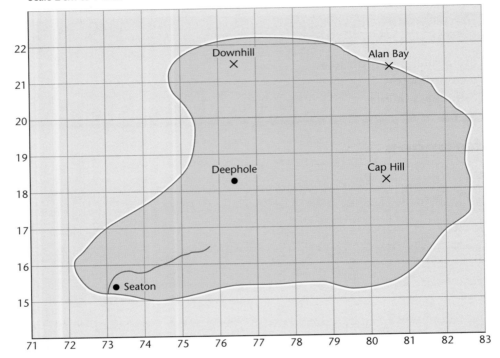

Scale 2 cm to 1 mile

a) Give the four-figure grid reference for **i)** Deephole **ii)** Seaton.

b) What is the direction from **i)** Deephole to Caphill, **ii)** Alan Bay to Caphill?

c) How far is it from **i)** Deephole to Caphill, **ii)** Alan Bay to Caphill?

a) **i)** 7618   **ii)** 7315

b) **i)** East   **ii)** South

c) **i)** 2 miles   **ii)** 1·5 miles

EXERCISE 6.1A

This is a sketch map of a village.

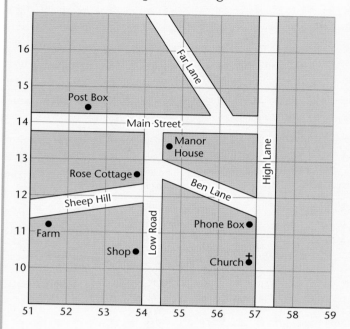

Answer these questions for this village.

1   Give the four-figure grid references for **a)** post box **b)** farm **c)** phone box.
2   Jane walked along Main Street towards High Lane. What direction did she walk?
3   Isaac walked north on Low Road from the shop. What is the name of the first road he passes on his left?
4   Steve left the farm on Sheep Hill to go to the church. Describe the route he should take.
5   What is the approximate direction that Far Lane runs from the Main Street?

Exercise 6.1A cont'd

This is a map of part of Cornwall.

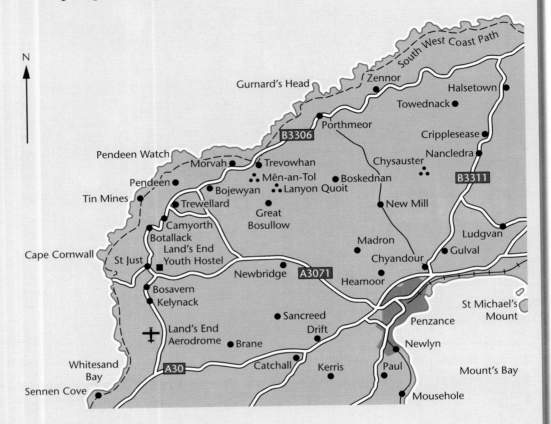

Answer these questions for the map.

6 Pam drove along the A3071 in a roughly westerly direction from Penzance. Name the first town that she passes close to on her left.

7 Fiona drove along the B3306 from Morvah to Porthmeor. Roughly in what direction did she drive?

8 Name the town roughly east of Great Bosullow.

9 The scale of the map is 1 cm to 2 km. Work out the straight line distance between

   a) St Just and Newbridge   **b)** Newbridge and Trevowhan.

10 Describe how to get from Trevowhan to Drift, using 'left' and 'right' at turnings and using the scale to give approximate distances along each road.

EXERCISE 6.1B

This is a map of part of Carlisle.

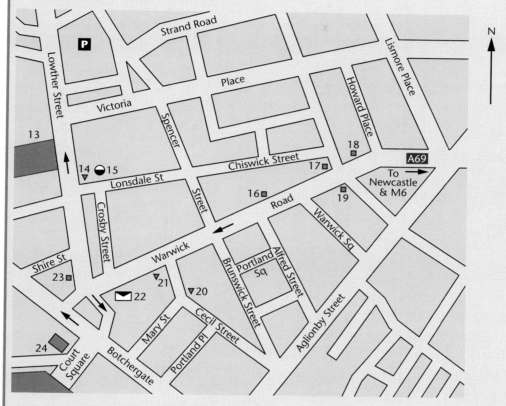

Answer these questions for the map.

1 Name a road that is roughly south-east to north-west in direction.
2 Paddy rode his bike north-east along Aglionby Street. What was the next road he passed on his left after Brunswick Street?
3 Beryl walked from the building 18 on Warwick Road away from Howard Place. What is the second road she passes on her right?
4 Debbie went from Lonsdale Street to Chiswick Street. Roughly what direction was that?
5 Describe how to get from the station at 15 on Lonsdale Street to the building at 16 on Warwick Road.

Chapter 6   *Maps, bearings and scale drawings*

Exercise 6.1B cont'd

This is a sketch map of an island.

Scale = 1 cm to 5 km

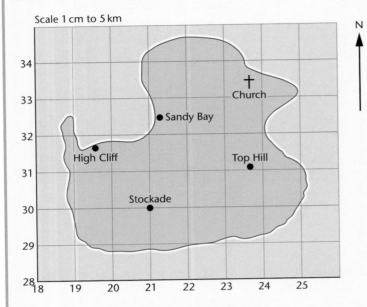

Answer these questions for the map.

6 Give the four-figure reference for
  a) Church   **b)** Top Hill   **c)** High Cliff.

7 What direction is it roughly from Sandy Bay to Top Hill?

8 What is north of Top Hill?

9 Work out the distance between **a)** Stockade and Top Hill.
  b) Top Hill and Sandy Bay.

10 Pam walks directly from Stockade to Top Hill to Sandy Bay, then round
  the coast to High Cliff and directly back to Stockade. How far does she
  walk altogether?

Bearings

If you want to describe the direction in which something is
moving, relative to your own position, you might say, 'Due
north' or 'north-east'. If you need to describe the direction
accurately you should use **bearings**. These describe a
direction as an angle measured clockwise from north.

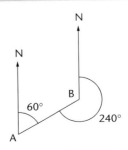

The most common type of bearing is a **three-figure bearing**. The three-figure bearing for 'due east' is 090°. The zero is put in front of the 90, to make up the three figures. A bearing of 'due south' is 180° and a bearing of 'due west' is 270°.

In the diagram on the previous page AB is on a bearing of 060° from A.

A is the fixed reference point for the bearing. Bearings are always given from some fixed point, such as a lighthouse.

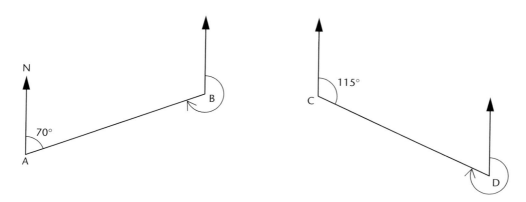

The bearing from A to B is 070°. Measure the bearing from B to A. This is the reverse bearing.

The bearing from C to D is 115°. Measure the bearing from C to D.

Draw other bearings from one point to another and measure the reverse bearing.

See if you can find a connection between each bearing and its reverse.

Exam tip

Bearings must have three figures so if the angle is less than 100° a zero must be put in front of the figures.
The bearing of A from B = the bearing of B from A + 180°
(or − 180° if the first bearing is more than 180°).

As the bearing is measured from A, the protractor is placed with its centre over A and its O on the north line.

You can write this as 'the bearing of B from A is 070°' or 'the bearing from A to B is 070°'.

You can also say 'the line BA is on a bearing of 250° from B' or 'the bearing of A from B is 250°' or 'the bearing from B of A is 250°'.

Chapter 6   *Maps, bearings and scale drawings*

EXAMPLE 2

Use a protractor to measure and write down the bearings of:

a) A from O

b) B from O

c) C from O.

a) 070°

b) 146°

c) 240°

If you have a circular protractor you can measure all of these directly.
If you only have a semicircular protractor then for **c)** you need to
measure the obtuse angle and subtract it from 360°.

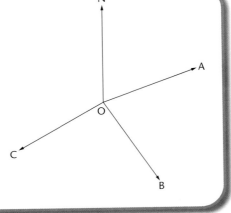

EXAMPLE 3

Three towns are Ayesford (A), Bagshot (B) and Carhill (C).

B is 20 km from A, on a bearing of 085° from A.

C is 15 km from B, on a bearing of 150° from B.

a) Make a scale drawing showing the three towns.
Use a scale of 1 cm to 5 km.

b) i) How far is Ayesford from Carhill?

ii) What is the bearing of Ayesford from Carhill?

a) Make a sketch and label the angles and lengths for the final diagram. Then draw
the diagram, starting far enough down the page to make sure it will all fit in.

Sketch   **Scale drawing (half-size)**

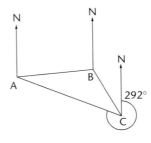

b) i) CA measures 5·9 cm, so the distance is 5·9 × 5 = 29·5 km.

ii) The bearing is 292°.

EXAMPLE 4

This is a sketch, not to scale, of three points in a field.

Work out the bearings of:

a) B from A   **b)** C from B   **c)** A from C.

Sketch

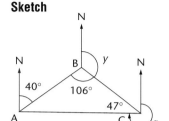

a) 040°.

b) The bearing of C from B is labelled y and the bearing of A from C is labelled q. From the diagram, the bearing of A from B is $y + 106°$. By calculation, the bearing of A from B is $180° + 40° = 220°$. So $y + 106° = 220°$

$y = 220° - 106° = 114°$.

The bearing of C from B = 114°.

c) From the diagram, the bearing of B from C is $q + 47°$. By calculation, the bearing of B from C = $180° + 114° = 294°$.

So $q + 47° = 294°$

$q = 294° - 47° = 247°$.

The bearing of A from C is 247°.

EXERCISE 6.2A

1 Measure the bearings of A, B, C and D from O.

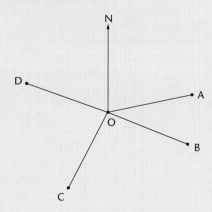

2 A boat is sailing due west. On what bearing is it sailing?

3 Tadmouth is on a bearing of 124° from Easton. What is the bearing of Easton from Tadmouth?

4 The road AB goes exactly north-east. On what bearing does it lie?

Exercise 6.2A cont'd

5 The map shows a lighthouse L and two boats A and B.

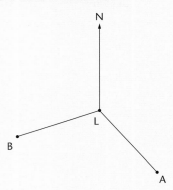

Measure the bearing from the lighthouse of:

a) boat A   **b)** boat B.

6 Dorton (D) is 8 miles from Brenwood (B), on a bearing of 058°. Make a sketch showing these two towns. Mark the distance and the angle clearly. Do not draw an accurate plan.

7 A boat leaves port and sails for 10 miles on a bearing of 285°. Make a sketch showing the direction in which the boat is sailing. Mark the distance and the angle clearly. Do not draw an accurate plan.

8 This sketch map shows the position of three oil platforms.

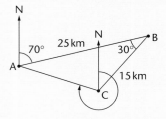

a) Make an accurate drawing of the sketch. Use a scale of 1 cm to 5 km.

b) Measure the bearing of A from C.

9 In an orienteering competition, competitors have to go from point A to point B to point C. A to B is 1 km on a bearing of 125°. B to C is 1·5 km on a bearing of 250°.

a) Draw an accurate plan of the route. Use a scale of 5 cm to 1 km.

b) Measure:

i) the bearing of C from A.
ii) the distance from C to A.

10 This sketch map shows the positions of three hills P, Q and R.

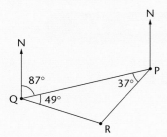

Work out the bearing of:

a) R from Q
b) R from P
c) P from R.

Do not make a scale drawing.

EXERCISE 6.2B

1 Mark a point O. Draw a north line through O. Mark points A, B and C, all 5 cm from O on these bearings.

OA 055°, OB 189°, OC 295°

2 Gravesend is on a bearing of 290° from Gillingham. What is the bearing of Gillingham from Gravesend?

3 An aeroplane is flying directly south-east. On what bearing is it flying?

4 On this drawing measure the bearing of:

a) P from Q   **b)** Q from R.

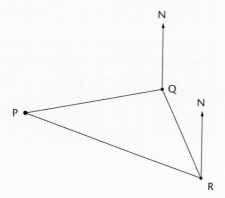

5 The sketch shows the position of three buoys in a harbour. Work out the bearing of:

a) A from B   **b)** C from A

Do not make a scale drawing.

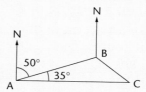

6 Kings Lynn is 60 km from Norwich, on a bearing from Norwich of 277°. Use a scale of 1 cm to 10 km to make an accurate drawing of the positions of these two towns.

7 This is an accurate drawing of the course of a yacht race.

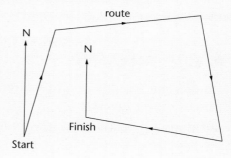

a) On what bearing do the yachts go on the first leg?

b) What is the bearing of the start from the finish?

Exercise 6.2B cont'd

8 David and Phil both leave point A and walk across the moors. After an hour David is 4 km from A on a bearing of 075°, and Phil is 5·5 km from A, on a bearing of 230°.

  a) Make an accurate drawing of their routes. Use a scale of 1 cm to 1 km.

  b) Find the bearing and distance of David from Phil.

  Remember to do a sketch first.

9 A glider leaves Sutton Bank airfield and flies 15 miles north-east. It then flies 22 miles east.

  a) Make a scale drawing of its journey. Use a scale of 1 cm to 5 miles.

  b) How far must it fly back to Sutton Bank, and on what bearing?

10 On the opposite sides of the entrance to a harbour are two radar beacons. The distance between them is 3·5 km and Beacon A is on a bearing of 110° from Beacon B. A ship is on a bearing of 320° from Beacon A and a bearing of 080° from Beacon B.

  Draw a map to a scale of 2 cm to 1 km, showing the two beacons and the ship.

Key ideas

- To find the four-figure grid reference of a place, take the left-hand bottom corner of the square.

- Bearings are measured clockwise from North and always have three figures.

7 Estimation

You should already know

- the measurements of familiar objects
- metric and imperial measures.

Estimating measurements

Sometimes you may be asked to estimate measures that you are not familiar with. For example, you may know how much a bag of sugar weighs, but what about a hen's egg? Would it be 5 g, 50 g or 500 g?

Here are some ideas.

> **Exam tip**
>
> So that you can estimate unfamiliar shapes and sizes, learn the measurements of a range of familiar ones.

- Know your own height, in metric and imperial units, such as 5 ft 7 in or 170 cm.
- Know how much you weigh, in metric and imperial units, such as 9 st 10 lb or 62 kg.
- Occasionally pick up a kilogram bag of sugar to remind yourself how heavy it feels.
- Measure your handspan. Knowing what it is (for example, 20 cm) enables you to measure the width of a table quickly in handspans, for instance, and then estimate the width in centimetres.
- Draw a 10 cm line and look at it carefully, to see how long it is. Practise drawing lines of given lengths without measuring them and see how close you can get to the correct length.
- For a 100 metre distance, think of the length of the 100 m race on an athletics track.

> **Exam tip**
>
> When estimating measures in unfamiliar contexts, try to compare them with measures you do know.

When comparing, ask yourself questions such as:

- Is it the same as …?
- Is it twice as long as …?
- Is it much heavier than …?
- How many of these would weigh the same as …?
- How many of these laid end to end would be the same length as …?

EXAMPLE 1

Choose the most suitable value, from this list, for the mass of the telephone directory for Guildford and West Surrey.

10 g   100 g   1000 g   10 000 g

A telephone directory is quite a large book. A mass of 10 g is quite small. Change the largest masses in the list to kilograms.

1000 g = 1 kg

10 000 g = 10 kg

A mass of 10 kg is too much to carry easily, but 1 kg is about right.

Checking with the smaller masses, 100 g is a small pack of cheese, for instance, and this would be too light in comparison.

So the answer is 1000 g.

Exam tip

When estimating angles remember that a square corner is 90° and a straight line is 180°.

EXERCISE 7.1A

1 Estimate the length of this line.

_____

2 Estimate the size of this angle.

3 Estimate the height of this tower.

Exercise 7.1A cont'd

4 Which of these is closest to the mass of a tablespoon of sugar?

2·5 g   25 g   250 g

5 Which of the masses below is likely to be how much Sarah is carrying in one of these shopping bags?

40 g   400 g   4 kg   40 kg

6 Estimate the size of angles A, B and C in this triangle.

7 Estimate the height of this building.

8 Estimate the lengths of the sides of the triangle.

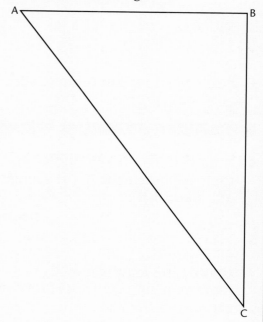

9 Estimate the amount of liquid in a teaspoon.

10 Estimate the mass of a 2 litre bottle of lemonade.

EXERCISE 7.1B

1 Estimate the length of this line.

2 Estimate the size of this angle.

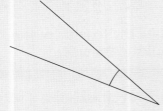

3 Estimate the length of this trailer.

4 Choose the most appropriate of these measurements for the length of a sports hall.

7 m   70 m   700 m   7000 m

5 When a large jug is full, it holds enough water to fill six tumblers. Choose the most likely measurement from the list for the capacity of the jug.

0·02 litres   0·2 litres

2 litres   20 litres

6

Estimate the height of the Christmas tree.

7

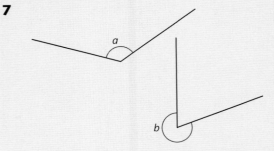

Estimate the size of angles *a* and *b*.

Exercise 7.1B cont'd

8 Estimate in kilometres how far you can walk in an hour.

9

Estimate the mass in kilograms that the lift will carry if the maximum number of people are in the lift.

10 Estimate the length of your longest finger.

Estimating area

When shapes are drawn on a grid the area can be estimated by counting squares. First count all the full squares, then make an estimate of the rest of the area by counting bits that are above half a square as a square and ignoring bits that are less than a square. If some bits are exactly half a square combine each 2 to make 1 square.

If the area you are finding is a scale drawing and you know what each square stands for you can find the area of the real place.

If in the example above each square stood for $4\,km^2$, then the area would be 14 or 15×4 which is between 56 and $60\,km^2$.

EXAMPLE 2

Estimate the area of this shape by counting squares.

There are 8 full squares, (marked F), 5 larger than half a square (marked L) and 3 half squares (marked H).

This gives $8 + 5 + 1\frac{1}{2} = 14\frac{1}{2}$. Therefore a sensible answer would be 14 or 15 squares.

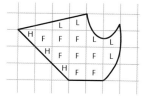

Exam tip

When estimating area, do not try to be too accurate. In example 2 any answer between 13 and 16 would be acceptable.

EXERCISE 7.2A

Estimate the area of these shapes by counting squares.

1

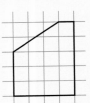

2

3

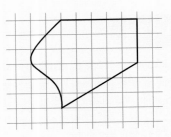

4

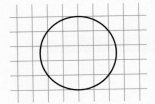

5

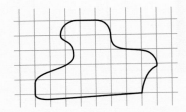

6 This is a map of the area of Folkestone and district. Each square is $1\,km^2$.

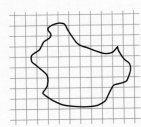

Estimate the area.

Exercise 7.2A cont'd

7 This is a map of Filey and district. Each square is 0·5 km².

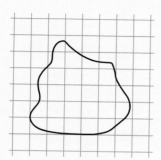

Estimate the area.

8 This is a map of a lake. Each square is 100 km².

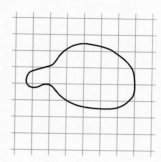

Estimate the area of the lake in m².

Estimate the area of these shapes by counting squares.

1

2

3

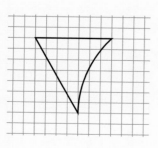

4

5

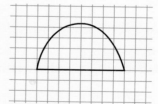

Exercise 7.2B cont'd

6 This is a map of East Anglia. Each square is 100 km^2.

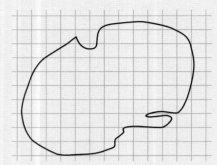

Estimate the area.

7 This is a plan of Zoe's house and garden. Each square is 2 m^2.

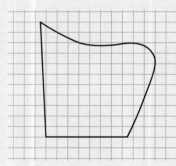

Estimate the area of her house and garden.

8 This is a plan of the front of Mr Frost's house that is to be paved. Each square is 2000 cm^2.

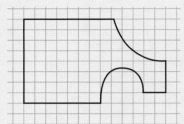

a) Estimate the area in cm^2.

b) It cost £30 per m^2 to have an area paved. How much will it cost Mr Frost?

Key ideas

● When estimating measures, compare with a measure you do know.
● When estimating areas, count the full squares and then estimate the rest from the parts of squares.

8 Pie charts

You should already know

- how to read a scale
- how to use a protractor.

Pie charts give a useful, clear picture and allow you to make comparisons between different types of data. They are often seen in newspapers.

Pie charts can be drawn quite easily using either a protractor or a pie chart scale.

ACTIVITY 1

Use a protractor to make an accurate copy of this circle, split into sectors.

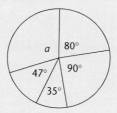

Measure angle *a* and make sure all the angles add up to 360°.

Look at a pie chart scale. It is like a protractor but instead of 360°, there are 100 divisions in the full circle. This means that each division is $\frac{1}{100}$ or 1 per cent.

ACTIVITY 2

Use a pie chart scale to make a copy of this circle, split into sectors.

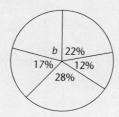

Measure the percentage *b* and make sure all the percentages add up to 100.

These examples explain how to draw a pie chart, using either a protractor or a pie chart.

EXAMPLE 1

Julie does a survey of 20 students in her class to find out which topic of mathematics they prefer. Draw a pie chart to show her results.

These are her results.

Topic	Number of students
Number work	8
Algebra	2
Geometry	4
Handling data	6

Answer

a) Using a pie chart scale

The total of 20 students = 100% on the scale

So 1 student = $\frac{100}{20}$ = 5%

This gives

Number work	8 × 5 =	40%
Algebra	2 × 5 =	10%
Geometry	4 × 5 =	20%
Handling data	6 × 5 =	30%
Total		100%

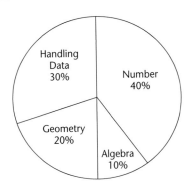

b) Using a protractor

The total of 20 students = 360°

So 1 student = $\frac{360}{20}$ = 18°

This gives

Number work	8 × 18 =	144°
Algebra	2 × 18 =	36°
Geometry	4 × 18 =	72°
Handling data	6 × 18 =	108°
Total		360°

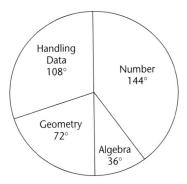

For the pie chart to mean anything it is vital that it is clearly labelled or a clear key is used, but the percentages or degrees are not needed.

Pie charts can be drawn for group data.

EXAMPLE 2

Draw the pie chart for the data in the table below. It shows the scores of the 25 students of class 6 in test A.

a) Using a pie chart scale   **b)** using a protractor.

Answer

a) As there are 25 students 1 student = $\frac{100}{25}$ = 4%

b) As there are 25 students 1 student = $\frac{360}{25}°$

As 360 ÷ 25 is not exact it is better not to work it out at this stage but to use a calculator and find each angle either exactly or to one decimal place.

Mark	Frequency	Percentage	Angle
0–9	4	4 × 4 = 16%	4 × $\frac{360}{25}$ = 57·6°
10–19	10	10 × 4 = 40%	10 × $\frac{360}{25}$ = 144°
20–29	8	8 × 4 = 32%	8 × $\frac{360}{25}$ = 115·2°
30–39	3	3 × 4 = 12%	3 × $\frac{360}{25}$ = 43·2°
Total	25	100%	360°

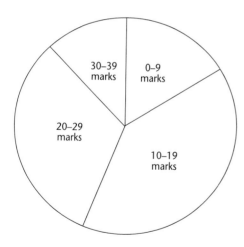

Using either angles or the percentages gives the same chart.

All the pie charts in the exercises can be drawn using either a pie chart scale or a protractor.

> ### Exam tip
> Work out the angle or percentage exactly or to one decimal place. Draw the angles or percentages as accurately as you can, but it is difficult to be more accurate than 1° or 1%. It is easier to work out the percentages or degrees in a column in the table. An important check is to make sure that the percentages add up to 100 or the degrees to 360.

EXERCISE 8.1A

1 Hannah asked 20 of her friends which was their favourite football team. There are her results. Draw a pie chart to illustrate her data.

Team	Number
Manchester United	9
Arsenal	5
Liverpool	3
Leeds	2
Chelsea	1
Total	20

2 This data gives the make of washing powder used by a sample of 200 housewives. Draw a pie chart to illustrate this data.

Make used	Number
Bold	16
Persil	38
Sainsbury	46
Tesco	60
Safeway	40
Total	200

3 At the Oasis cafe, Cyril kept a list of the drinks ordered in one day.

Drink	Number
Tea	70
Coffee	50
Hot chocolate	26
Diet coke	18
Apple juice	11
Coke	25
Total	200

The results are shown in the table. Draw a pie chart to illustrate this data.

4 The lengths of 50 pea pods are given in this table.

Draw a pie chart showing the lengths of the pea pods.

Length (mm)	Number
50–59	4
60–69	18
70–79	12
80–89	10
90–99	6
Total	50

Exercise 8.1A cont'd

5

– 2metres

Students in year 10 were asked to estimate the height of a mark 2 metres up a wall. Here are their estimates in centimetres.

194	180	212	196	182	186	184	156	220	175
180	186	156	158	162	192	184	218	204	164
214	160	160	198	204	188	184	184	190	206
202	198	184	179	169	178	192	215	208	202

Group this data into intervals 151–160 etc. and show it in a pie chart.

6 Tom earns £20 each week, washing cars at the local garage.

He divides his earnings as follows:

Savings £5·00

Clothes £7·00

Cds etc. £4·50

Going out £3·50.

Draw a pie chart to illustrate this.

7 The nutritional information on a packet of breakfast cereal states

Draw a pie chart to show this.

Typical value per 100g			
Protein	6 g	Sodium	1·1 g
Sugar	10 g	Fibre	1·5 g
Starch	75 g	Other	5·1 g
Fat	1·3 g		

8 Make a list of five popular songs and ask 25 of the students in your class to pick their favourite. Draw a pie chart to show the information you have collected.

EXERCISE 8.1B

1 Sanjay did a survey of 50 cars and checked their makes. These are his results.

Ford 15   Vauxhall 10   Toyota 20   VW 5.

Draw a pie chart of this information.

2 Matilda asked 25 of her classmates which of five TV soaps they preferred.

These are here results.

Eastenders 5   Coronation Street 6   Emmerdale 4   Neighbours 8
Brookside 2.

Draw a pie chart of these results.

3 The table shows the approximate 1998 population (in millions) for six African countries. Show this information in a pie chart.

Country	Population (millions)
Rwanda	8
Somalia	10·6
Uganda	21
Tanzania	30·4
Nigeria	121·6
Ethiopia	58·4
Total	250

4 Look in a newspaper or look round the class you are in. Take 15 male names and 15 female names. Draw pie charts to show the frequency of the names of males and females. Compare the two pie charts.

5 It cost Mr and Mrs Jones £800 to go on holiday with their children.

The costs were split as follows.

Caravan £350   Transport £80   Food £200   Spending £170.

Draw a pie chart to illustrate this information.

6 An hospital listed the ages of 200 women who had babies in its maternity ward during the first two months of the year. Draw a pie chart of this information.

Age range	Number of women
16–20	15
21–25	49
26–30	71
31–35	44
36–40	16
41–45	3
46–50	2
Total	200

Exercise 8.1B cont'd

7 The daily choice of newspaper at the 50 houses on Peter's paper round are:

Daily Mail 7   Express 4   Telegraph 8   Guardian 9
Sun   13   Mirror 9.

Draw a pie chart to illustrate this.

8 The salaries of 80 men are given in this table.

Draw a pie chart to illustrate these salaries.

Salary (£1000)	Number
$10 \leqslant x < 15$	16
$15 \leqslant x < 20$	12
$20 \leqslant x < 25$	20
$25 \leqslant x < 30$	10
$30 \leqslant x < 35$	8
$35 \leqslant x < 40$	14
Total	80

Information from pie charts

Again either a pie chart scale or a protractor can be used.

EXAMPLE 3

Harry earns £24 000 a year. This pie chart shows how he uses the money.

How much does he spend on **a)** food, **b)** travel?

Answers

i) Pie chart scale 1% = £24 000 ÷ 100 = 240.

   a) Food   = 20 × 240 = £4800

   b) Travel   = 15 × 240 = £3600

> Food sector is 20%
> Travel sector is 15%

ii) Protractor 1° = 24 000 ÷ 360

   a) Food   = 72 × 24 000 ÷ 360 = £4800

   b) Travel   = 54 × 24 000 ÷ 360 = £3600

> Food sector is 72°
> Travel sector is 54°

When reading information from a pie chart, first find what 1% or 1° equals. As in Example 3 you need not work out the value of 1°

Exam tip

When measuring from a pie chart do not try to be more accurate than 1 decimal place and often the nearest degree or percentage is as accurate as you can be.

EXERCISE 8.2A

1

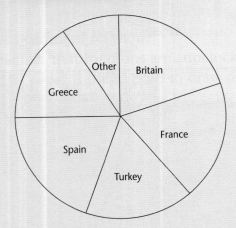

2

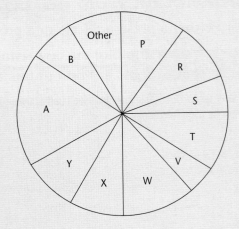

This pie chart shows the holiday destinations of the 100 students in year 9 of Park Ridge school.

How many pupils had their holiday in **a)** Britain **b)** Greece?

This pie chart shows the year letter of cars in the school car park. There were 200 cars altogether.

a) What was the most common letter?

b) How many cars had letter **(i)** P **(ii)** V?

Exercise 8.2A cont'd

3

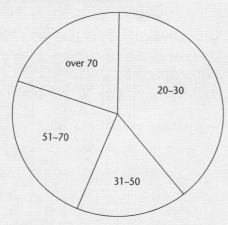

This pie chart shows the age groups of the first 100 people through the doors of a new supermarket. How many were there in each age group?

4

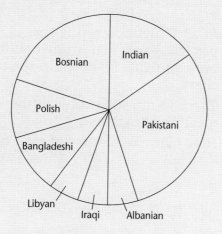

This pie chart shows the country of birth of 60 adults attending an English class. How many were there from each country?

5 a) In a survey for a by-election in Scotland 200 people were asked their voting intentions.

These are the results.

Labour	52	Conservative	40
Lib-Dem	30	S.N.P.	64
Don't know	14		

Draw a pie chart of this information.

b) In the actual election, the proportion who voted for each party is shown in the pie chart.

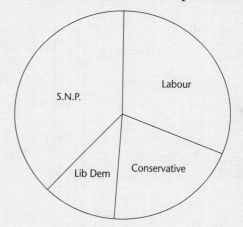

Compare the survey results with the actual results.

6 See if you can find some pie charts in newspapers or magazines. Cut them out and analyse what they show.

EXERCISE 8.2B

1 One thousand people were interviewed about their holiday destination. This pie chart shows this information.

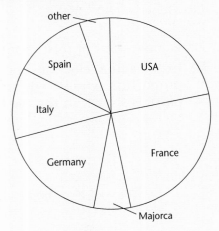

Work out how many went to each country.

2

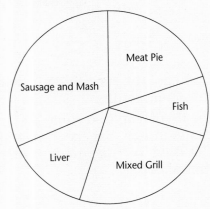

This pie chart shows the type of meal ordered at Joe's cafe during a day.

a) Which was the most popular choice.

b) There were 80 customers. How many ordered

(i) Meat pie   (ii) Mixed grill?

3

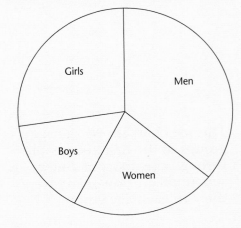

At an athletics meeting the percentage of men, women and boys and girls is shown in the pie chart.

There were 7500 people at the meeting altogether.

How many were there of each type?

Exercise 8.2B cont'd

4

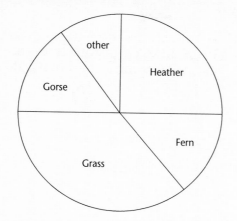

The area covered by different types of plants growing on a 1000 m² section of moorland. Find the area covered by each plant.

5

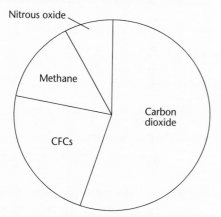

An analysis of greenhouse gases in the atmosphere is shown in the pie chart. What is the percentage of each gas present in the atmosphere?

6 a) In their mock exams the 25 members of 11A maths set obtained the following grades:

A 10   B 7   C 6   D 1   E 1.

Draw a pie chart of these grades.

b) In the final exam this pie chart shows the grades of the same 25 students.

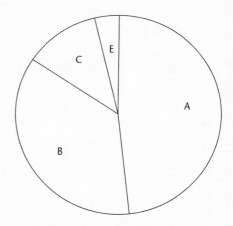

Compare the mock and final grades.

Key ideas

● To draw a pie chart work out what percentage or angle represents each item of data and use the pie chart scale or protractor to draw it.

● To find how many items are represented by each sector of a pie chart, work out what 1% or 1° stands for and multiply each reading from the pie chart by this amount.

B1 Revision exercise

1 Write down the answers to these calculations.

a) 7^2 **b)** 11^2
c) $5^2 + 6^2$ **d)** $1^2 + 2^2 + 3^2$
e) $\sqrt{144}$ **f)** $\sqrt{169}$
g) $\sqrt{5^2 - 4^2}$ **h)** $\sqrt[3]{27}$.

2 Investigate the last digits in the squares of the whole numbers from 1 to 20.

Using these results, decide which of these numbers are definitely not square numbers.

17178 17170 17179 17172 17177

3 Write down all the factors of
a) 6 **b)** 15 **c)** 24 **d)** 30 **e)** 64.

4

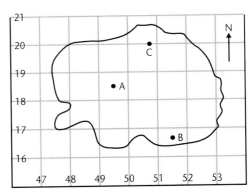

a) Write down the four-figure grid reference of **i)** A, **ii)** B.
b) Lucy walked directly from A to B. What approximate direction did she walk?
c) Tom walked south from C. Was A on his left or his right?

5 This is a sketch of four trees in a field, with some angles marked.

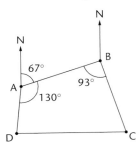

Work out the bearing of
a) A from B **b)** C from B
c) A from D.

6 This is a sketch of three buoys in the sea.

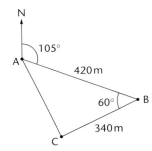

Make a scale drawing and find the distance and bearing of A from C. Use a scale of 1 cm to 50 m.

7 Theresa is taking part in an yacht race. She sails 15 miles on a bearing of 140°, then 12 miles on a bearing of 025°. Use a scale of 1 cm to 2 miles to make a scale drawing of her route. Find how far she is from the start and on what bearing.

8 Estimate the height of this work-top.

9 Walking at a normal pace, it takes Yasmin five minutes to walk from home to school. Estimate how far her school is from her home.

10 Estimate the following using metric units.

  a) the length of a swimming pool
  b) the length of a chair leg
  c) the mass of a jar of jam
  d) the capacity of a soup bowl
  e) the capacity of a kitchen sink
  f) the mass of an egg
  g) the length of the bristles in a tooth brush.

11 Estimate the size of angles *a*, *b* and *c*.

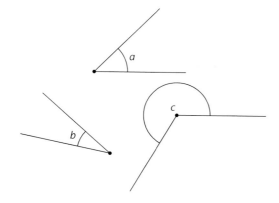

12

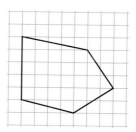

Estimate the area of the polygon.

13 During a year a grain merchant sells the following numbers of sacks of grain.

Barley	12 000	Maize	9000
Corn	15 000	Wheat	24 000

Show this information on a pie chart.

14 Tariq did a survey about the lunch-time eating arrangements of his year group. The pie chart shows his findings.

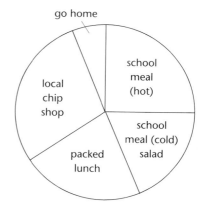

If 97 children have the hot school meal, calculate the numbers for the other choices.

9 Fractions

You should already know

- how to multiply and divide by simple numbers.

Fractions

In a fraction $\dfrac{a}{b}$, a is called the numerator and b is the denominator.

Equivalent fractions

These squares can be divided into equal parts in different ways.
The fraction represented by the shaded parts can be thought of as $\frac{1}{4}$ or $\frac{2}{8}$ or $\frac{4}{16}$.

These three fractions are equal in value and are equivalent.

ACTIVITY 1

a) On a piece of squared paper draw a rectangle 3 by 4 squares.
  i) Use this as a whole one and shade in 3 squares of the shape.
  ii) Write this as a fraction with denominator 12.
  iii) Find another fraction that is equivalent to this.
b) Draw other rectangles and squares and find equivalent fractions in them.

You can find equivalent fractions by multiplying or dividing the numerator *and* the denominator by the same number.

EXAMPLE 1

Fill in the missing numbers in these equivalent fractions.

a) $\frac{1}{2} = \frac{2}{\square} = \frac{\square}{14}$ **b)** $\frac{3}{8} = \frac{\square}{24} = \frac{15}{\square}$

a) $\frac{1}{2} = \frac{2}{4} = \frac{7}{14}$

Multiply the first fraction by $\frac{2}{2}$, then multiply it by $\frac{7}{7}$.

b) $\frac{3}{8} = \frac{9}{24} = \frac{15}{40}$

Multiply the first fraction by $\frac{3}{3}$, then multiply it by $\frac{5}{5}$.

You may be asked to write a fraction in its lowest terms.
This means finding the smallest possible denominator.

This is sometimes called cancelling the fractions.

EXAMPLE 2

Write these fractions in their lowest terms.

a) $\frac{6}{10}$ **b)** $\frac{8}{12}$ **c)** $\frac{15}{20}$.

a) $\frac{6}{10} = \frac{3}{5}$

Divide the numerator and denominator by 2.

b) $\frac{8}{12} = \frac{2}{3}$

$\frac{8}{12}$ is also equivalent to $\frac{4}{6}$ but it can be simplified further to $\frac{2}{3}$.

c) $\frac{15}{20} = \frac{3}{4}$

Divide top and bottom by 5.

Ordering fractions

To put fractions in order, change them to the same denominator and order
by the numerator.

EXAMPLE 3

Put these fractions in order, smallest first.

$\frac{3}{10}$ $\frac{1}{4}$ $\frac{9}{20}$ $\frac{2}{5}$ $\frac{1}{2}$

All these denominators divide into 20 so they can all be changed to a denominator of 20.

They become

$\frac{6}{20}$ $\frac{5}{20}$ $\frac{9}{20}$ $\frac{8}{20}$ $\frac{10}{20}$

So order is $\frac{5}{20}$ $\frac{6}{20}$ $\frac{8}{20}$ $\frac{9}{20}$ $\frac{10}{20}$

or $\frac{1}{4}$ $\frac{3}{10}$ $\frac{2}{5}$ $\frac{9}{20}$ $\frac{1}{2}$

EXERCISE 9.1A

1 Fill in the blanks in these equivalent fractions.

$\frac{1}{4} = \frac{}{8} = \frac{}{12} = \frac{5}{}$

2 Fill in the blanks in these equivalent fractions.

$\frac{2}{5} = \frac{4}{} = \frac{}{25} = \frac{12}{}$

3 Write these fractions in their lowest terms.

a) $\frac{6}{9}$ b) $\frac{15}{25}$ c) $\frac{4}{12}$ d) $\frac{18}{27}$

4 Write these fractions in order, smallest first.

$\frac{7}{10}$ $\frac{3}{4}$ $\frac{11}{20}$ $\frac{3}{5}$

5 Write these fractions in order, smallest first.

$\frac{13}{15}$ $\frac{2}{3}$ $\frac{3}{10}$ $\frac{2}{5}$ $\frac{1}{2}$

EXERCISE 9.1B

1 Fill in the blanks in these equivalent fractions.

$\frac{1}{5} = \frac{}{10} = \frac{}{20} = \frac{7}{}$

2 Fill in the blanks in these equivalent fractions.

$\frac{2}{9} = \frac{4}{} = \frac{}{36} = \frac{6}{}$

3 Write these fractions in their lowest terms.

a) $\frac{6}{16}$ b) $\frac{15}{35}$ c) $\frac{12}{18}$ d) $\frac{28}{42}$

4 Write these fractions in order, smallest first.

$\frac{7}{12}$ $\frac{3}{4}$ $\frac{7}{8}$ $\frac{5}{6}$

5 Write these fractions in order, smallest first.

$\frac{13}{16}$ $\frac{5}{8}$ $\frac{3}{4}$ $\frac{7}{16}$ $\frac{1}{2}$

Adding fractions

In this diagram, each square is divided into thirds.

The diagram shows $\frac{1}{3} + \frac{1}{3} = \frac{2}{3}$.

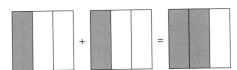

> Counting squares or columns, 1 column + 1 column = 2 columns. If the denominator is the same in the fractions, just add the numerators.

In this diagram there are 12 small squares in each rectangle.

The diagram shows $\frac{1}{3} + \frac{1}{4}$.

Counting squares, 4 squares + 3 squares = 7 squares or $\frac{7}{12}$.

> To add two fractions, change them to the same type, so they have the same denominator.

$\frac{1}{3} = \frac{4}{12}$ $\frac{1}{4} = \frac{3}{12}$ $\frac{4}{12} + \frac{3}{12} = \frac{7}{12}$

EXAMPLE 4

Add the fractions.

a) $\frac{2}{5} + \frac{1}{5}$ **b)** $\frac{1}{5} + \frac{3}{10}$ **c)** $\frac{1}{6} + \frac{3}{4}$

a) $\frac{2}{5} + \frac{1}{5} = \frac{3}{5}$ | They have the same denominator so just add the numerators.

b) $\frac{1}{5} + \frac{3}{10}$ | They need to be changed so that they both have the same denominator.

 $= \frac{2}{10} + \frac{3}{10}$ | Change $\frac{1}{5}$ to $\frac{2}{10}$ so they both have 10 as the denominator.

 $= \frac{5}{10} = \frac{1}{2}$ | Write the answer in the lowest terms.

c) $\frac{1}{6} + \frac{3}{4}$ | This time they both need to be changed, to have the same denominator.

 $= \frac{2}{12} + \frac{9}{12} = \frac{11}{12}$ $\frac{1}{6} = \frac{2}{12} = \frac{3}{18} = \dots, \frac{3}{4} = \frac{6}{8} = \frac{9}{12} \dots$

Subtracting fractions

In this diagram there are 20 small squares in each rectangle.

The diagram shows $\frac{1}{4} - \frac{1}{5}$

5 squares – 4 squares = 1 square = $\frac{1}{20}$

This can be written as $\frac{1}{4} - \frac{1}{5} = \frac{5}{20} - \frac{4}{20} = \frac{1}{20}$

Change both so that they have the same denominator, as with adding, but then subtract the numerators.

> **Exam tip**
>
> When adding or subtracting fractions, change to the same type (same denominator) and add or subtract the numerators. The most common error is to add the denominators and add the numerators.

Chapter 9   *Fractions*

EXAMPLE 5

Work these out.

a) $\frac{3}{5} - \frac{2}{5}$ **b)** $\frac{3}{4} - \frac{2}{3}$ **c)** $\frac{5}{6} + \frac{1}{4} - \frac{1}{3}$

a) $\frac{3}{5} - \frac{2}{5} = \frac{1}{5}$ They have the same denominator so just add the numerators.

b) $\frac{3}{4} - \frac{2}{3} = \frac{9}{12} - \frac{8}{12}$ 4 and 3 both divide into 12, so make 12 the denominator for both.

$\quad = \frac{1}{12}$ Multiply $\frac{3}{4}$ by 3 top and bottom, and $\frac{2}{3}$ by 4 top and bottom.

c) $\frac{5}{6} + \frac{1}{4} - \frac{1}{3} = \frac{10}{12} + \frac{3}{12} - \frac{4}{12}$ All the denominators divide into 12.

$\quad = \frac{9}{12}$

$\quad = \frac{3}{4}$

EXERCISE 9.2A

1 Fill in the blanks in these equivalent fractions.

   a) $\frac{1}{7} = \frac{2}{\square} = \frac{\square}{35}$

   b) $\frac{4}{9} = \frac{16}{\square} = \frac{\square}{72}$

2 Write these fractions in their lowest terms.

   a) $\frac{6}{8}$ **b)** $\frac{12}{15}$ **c)** $\frac{12}{24}$ **d)** $\frac{12}{54}$

   For the rest of the questions, give all answers in their lowest terms.

3 Add these fractions.

   a) $\frac{1}{8} + \frac{1}{8}$ **b)** $\frac{1}{8} + \frac{3}{8}$

   c) $\frac{1}{4} + \frac{1}{8}$ **d)** $\frac{1}{2} + \frac{3}{8}$

   e) $\frac{5}{8} + \frac{1}{4}$

4 Subtract these fractions.

   a) $\frac{3}{8} - \frac{1}{8}$ **b)** $\frac{5}{8} - \frac{3}{8}$

   c) $\frac{1}{4} - \frac{1}{8}$ **d)** $\frac{1}{2} - \frac{3}{8}$

   e) $\frac{5}{8} - \frac{1}{4}$

5 Add these fractions.

   a) $\frac{2}{3} + \frac{1}{3}$ **b)** $\frac{1}{3} + \frac{1}{2}$

   c) $\frac{3}{5} + \frac{1}{4}$ **d)** $\frac{1}{6} + \frac{2}{3}$

   e) $\frac{2}{5} + \frac{3}{8}$ **f)** $\frac{3}{4} + \frac{1}{6}$

6 Subtract these fractions.

   a) $\frac{2}{7} - \frac{1}{7}$ **b)** $\frac{5}{6} - \frac{1}{3}$

   c) $\frac{2}{3} - \frac{1}{4}$ **d)** $\frac{11}{12} - \frac{2}{3}$

   e) $\frac{5}{8} - \frac{1}{3}$ **f)** $\frac{7}{9} - \frac{5}{12}$

7 Add these fractions.

   a) $\frac{1}{8} + \frac{1}{8} + \frac{1}{8}$ **b)** $\frac{1}{8} + \frac{3}{8} + \frac{1}{4}$

   c) $\frac{1}{4} + \frac{1}{8} + \frac{1}{8}$ **d)** $\frac{1}{2} + \frac{1}{8} + \frac{3}{8}$

   e) $\frac{1}{8} + \frac{1}{4} + \frac{1}{8}$

8 Work these out.

   a) $\frac{4}{5} + \frac{7}{10} - \frac{3}{5}$ **b)** $\frac{3}{5} + \frac{5}{6} - \frac{2}{3}$

   c) $\frac{2}{3} + \frac{3}{4} - \frac{1}{2}$ **d)** $\frac{2}{5} + \frac{5}{8} - \frac{3}{4}$

   e) $\frac{1}{5} + \frac{3}{10} - \frac{1}{2}$ **f)** $\frac{3}{7} + \frac{5}{14} - \frac{1}{2}$

9 Write these fractions in order, smallest first.

$$\frac{1}{2} \quad \frac{1}{8} \quad \frac{3}{8} \quad \frac{1}{4}$$

10 Write these fractions in order, smallest first.

$$\frac{11}{12} \quad \frac{5}{8} \quad \frac{3}{4} \quad \frac{7}{24} \quad \frac{1}{2}$$

1 Fill in the blanks in these equivalent fractions.

a) $\frac{1}{6} = \frac{4}{\square} = \frac{\square}{12}$

b) $\frac{2}{3} = \frac{\square}{6} = \frac{12}{\square} = \frac{\square}{24}$

2 Write these fractions in their lowest terms.

a) $\frac{8}{16}$ **b)** $\frac{9}{15}$ **c)** $\frac{10}{25}$ **d)** $\frac{24}{30}$

For the rest of the questions, give all answers in their lowest terms.

3 Add these fractions.

a) $\frac{1}{10} + \frac{1}{10}$ **b)** $\frac{1}{10} + \frac{3}{10}$

c) $\frac{1}{5} + \frac{1}{10}$ **d)** $\frac{1}{2} + \frac{3}{10}$

e) $\frac{3}{10} + \frac{2}{5}$

4 Subtract these fractions.

a) $\frac{3}{10} - \frac{1}{10}$ **b)** $\frac{5}{10} - \frac{3}{10}$

c) $\frac{1}{5} - \frac{1}{10}$ **d)** $\frac{1}{2} - \frac{3}{10}$

e) $\frac{7}{10} - \frac{2}{5}$

5 Add these fractions.

a) $\frac{2}{7} + \frac{4}{7}$ **b)** $\frac{1}{3} + \frac{1}{6}$

c) $\frac{2}{3} + \frac{1}{4}$ **d)** $\frac{1}{5} + \frac{3}{4}$

e) $\frac{3}{8} + \frac{1}{5}$ **f)** $\frac{3}{4} + \frac{2}{5}$

6 Subtract these fractions.

a) $\frac{3}{4} - \frac{1}{4}$ **b)** $\frac{1}{2} - \frac{1}{3}$

c) $\frac{3}{4} - \frac{3}{5}$ **d)** $\frac{3}{4} - \frac{1}{6}$

e) $\frac{3}{5} - \frac{1}{2}$ **f)** $\frac{7}{8} - \frac{2}{3}$

7 Add these fractions.

a) $\frac{1}{12} + \frac{1}{12} + \frac{3}{12}$

b) $\frac{1}{12} + \frac{1}{4} + \frac{1}{3}$

c) $\frac{1}{4} + \frac{5}{12} + \frac{1}{6}$

d) $\frac{1}{10} + \frac{1}{5} + \frac{3}{10}$

e) $\frac{1}{5} + \frac{1}{20} + \frac{3}{10}$

8 Work these out.

a) $\frac{3}{5} + \frac{2}{5} - \frac{7}{10}$ **b)** $\frac{1}{4} + \frac{3}{8} - \frac{1}{6}$

c) $\frac{1}{6} + \frac{2}{3} - \frac{1}{4}$ **d)** $\frac{5}{8} + \frac{3}{5} - \frac{3}{4}$

e) $\frac{3}{4} - \frac{5}{6} + \frac{2}{3}$ **f)** $\frac{3}{20} - \frac{2}{5} + \frac{3}{4}$

9 Write these fractions in order, smallest first.

$\frac{1}{12}$ $\frac{1}{3}$ $\frac{3}{4}$ $\frac{1}{6}$ $\frac{1}{2}$

10 Write these fractions in order, smallest first.

$\frac{11}{16}$ $\frac{7}{8}$ $\frac{3}{4}$ $\frac{17}{32}$

Mixed numbers

Look at this calculation.

$\frac{2}{3} + \frac{2}{3} = \frac{4}{3}$

As you an see the result of this addition is a fraction which is 'top-heavy'. It is usual to write fractions like this as **mixed numbers**.

$\frac{4}{3} = 1\frac{1}{3}$

To change a top-heavy fraction to a mixed number, divide the denominator into the numerator and write the remainder as a fraction over the denominator.

EXAMPLE 6

Change these fractions to mixed numbers.

a) $\frac{7}{4}$ b) $\frac{11}{5}$ c) $\frac{24}{7}$

a) $\frac{7}{4} = 1\frac{3}{4}$ $7 \div 4 = 1$ with 3 left over.

b) $\frac{11}{5} = 2\frac{1}{5}$ $11 \div 5 = 2$ with 1 left over.

c) $\frac{24}{7} = 3\frac{3}{7}$ $24 \div 7 = 3$ with 3 left over.

Exam tip

The most common error is to put the remainder over the numerator rather than the denominator.

Mixed numbers can be changed to top-heavy fractions. Just reverse the process.

EXAMPLE 7

Change these mixed numbers to top-heavy fractions.

a) $3\frac{1}{4}$ b) $2\frac{3}{5}$ c) $3\frac{5}{6}$

a) $3\frac{1}{4} = 3 + \frac{1}{4} = \frac{12}{4} + \frac{1}{4} = \frac{13}{4}$

Change the whole number to quarters and then add. Another way to think of it is to multiply the whole number by the denominator and add on the numerator.
$(3 \times 4 + 1 = 13)$.

b) $2\frac{3}{5} = \frac{13}{5}$ $2 \times 5 + 3 = 13$

c) $3\frac{5}{6} = \frac{23}{6}$ $3 \times 6 + 5 = 23$

To add or subtract mixed numbers deal with the whole numbers first.

EXAMPLE 8

Work these out.

a) $1\frac{1}{6} + 2\frac{1}{3}$ b) $2\frac{3}{4} + \frac{3}{5}$ c) $3\frac{2}{3} - 1\frac{1}{6}$ d) $4\frac{1}{5} - 1\frac{1}{2}$

a) $1\frac{1}{6} + 2\frac{1}{3} = 3 + \frac{1}{6} + \frac{1}{3}$

$= 3 + \frac{1}{6} + \frac{2}{6}$

$= 3\frac{3}{6} = 3\frac{1}{2}$

Add the whole numbers, then deal with the fractions in the normal way.

b) $2\frac{3}{4} + \frac{3}{5} = 2 + \frac{15}{20} + \frac{12}{20}$

$= 2 + \frac{27}{20}$

$= 2 + 1 + \frac{7}{20}$

$= 3\frac{7}{20}$

You end up with a top-heavy fraction which you have to change to a mixed number, and then add the whole numbers.

Example 8 cont'd

c) $3\frac{2}{3} - 1\frac{1}{6} = 3 - 1 + \frac{2}{3} - \frac{1}{6}$

$= 2 + \frac{4}{6} - \frac{1}{6}$

$= 2\frac{3}{6} = 2\frac{1}{2}$

d) $4\frac{1}{5} - 1\frac{1}{2} = 3 + \frac{2}{10} - \frac{5}{10}$

$= 2 + \frac{10}{10} + \frac{2}{10} - \frac{5}{10}$

$= 2\frac{7}{10}$

Subtract the numbers and then the fractions.

Working out $\frac{2}{10} - \frac{5}{10}$ gives an answer of negative $\frac{3}{10}$.
Change one of the whole numbers into $\frac{10}{10}$, then subtract.

EXERCISE 9.3A

1 Change these top-heavy fractions to mixed numbers.

a) $\frac{11}{8}$ **b)** $\frac{15}{8}$ **c)** $\frac{9}{4}$
d) $\frac{7}{2}$ **e)** $\frac{15}{7}$

2 Change these mixed numbers to top-heavy fractions.

a) $1\frac{1}{8}$ **b)** $2\frac{5}{8}$ **c)** $3\frac{1}{4}$
d) $5\frac{1}{2}$ **e)** $5\frac{1}{5}$

3 Add, write your answers as simply as possible.

a) $\frac{3}{10} + 1\frac{2}{5}$ **b)** $1\frac{1}{10} + \frac{3}{5}$
c) $2\frac{1}{5} + 1\frac{1}{10}$ **d)** $4\frac{1}{2} + 2\frac{3}{10}$
e) $1\frac{3}{10} + \frac{2}{5}$

4 Subtract, write your answers as simply as possible.

a) $2\frac{3}{10} - \frac{1}{10}$ **b)** $3\frac{5}{10} - 1\frac{3}{10}$
c) $4\frac{1}{5} - 2\frac{1}{10}$ **d)** $3\frac{1}{2} - \frac{3}{10}$
e) $6\frac{7}{10} - 6\frac{2}{5}$

5 Add, write your answers as simply as possible.

a) $\frac{5}{12} + \frac{1}{12} + \frac{3}{4}$ **b)** $\frac{7}{12} + \frac{1}{4} + \frac{2}{3}$
c) $\frac{3}{4} + \frac{5}{12} + \frac{1}{6}$ **d)** $\frac{1}{12} + \frac{5}{6} + \frac{3}{4}$
e) $\frac{1}{5} + \frac{17}{20} + \frac{3}{10}$

6 Subtract, write your answers as simply as possible.

a) $1\frac{3}{8} - \frac{5}{8}$ **b)** $2\frac{5}{8} - 1\frac{3}{8}$
c) $3\frac{1}{4} - 1\frac{5}{8}$ **d)** $3\frac{1}{2} - 1\frac{7}{8}$
e) $3\frac{1}{8} - \frac{1}{4}$

7 Write down two fractions that add up to $\frac{3}{4}$.

8 Gurdeep had a packet of sweets. He ate $\frac{1}{2}$ of them himself and gave Torin $\frac{1}{8}$. What fraction did he have left.

EXERCISE 9.3B

1 Change these top-heavy fractions to mixed numbers.

   a) $\frac{11}{4}$ **b)** $\frac{5}{2}$ **c)** $\frac{9}{5}$

   d) $\frac{7}{3}$ **e)** $\frac{15}{8}$

2 Change these mixed numbers to top-heavy fractions.

   a) $1\frac{1}{2}$ **b)** $2\frac{2}{5}$ **c)** $3\frac{1}{3}$

   d) $5\frac{1}{4}$ **e)** $2\frac{3}{8}$

3 Add, write your answers as simply as possible.

   a) $1\frac{5}{8} + \frac{1}{8}$ **b)** $\frac{1}{8} + 2\frac{3}{8}$

   c) $1\frac{3}{4} + 2\frac{1}{8}$ **d)** $3\frac{1}{2} + \frac{7}{8}$

   e) $4\frac{5}{8} + \frac{3}{4}$

4 Subtract, write your answers as simply as possible.

   a) $1\frac{3}{8} - \frac{1}{8}$ **b)** $5\frac{5}{8} - 3\frac{3}{8}$

   c) $2\frac{1}{4} - 1\frac{1}{8}$ **d)** $4\frac{1}{2} - 3\frac{3}{8}$

   e) $5\frac{5}{8} - 1\frac{1}{4}$

5 Add, write your answers as simply as possible.

   a) $\frac{1}{8} + \frac{3}{8} + \frac{7}{8}$ **b)** $\frac{7}{8} + \frac{3}{8} + \frac{1}{4}$

   c) $\frac{3}{4} + 1\frac{1}{8} + \frac{1}{8}$ **d)** $1\frac{1}{2} + \frac{7}{8} + \frac{3}{4}$

   e) $2\frac{1}{2} + \frac{1}{4} + \frac{3}{8}$

6 Subtract, write your answers as simply as possible.

   a) $5\frac{1}{10} - \frac{7}{10}$ **b)** $1\frac{5}{6} - \frac{2}{3}$

   c) $1\frac{1}{5} - \frac{7}{10}$ **d)** $1\frac{1}{2} - \frac{3}{4}$

   e) $2\frac{7}{10} - 1\frac{4}{5}$

7 Put these fractions in order, smallest first.

   $\frac{5}{2}$ $1\frac{2}{5}$ $1\frac{1}{3}$ $1\frac{1}{4}$ $\frac{11}{8}$

8 Sophie went on a 3-day walk. She walked $\frac{1}{2}$ the distance on the first day and $\frac{3}{10}$ of the distance on the second day. What fraction was left to be covered on the third day?

Key ideas

- To order fractions, change to the same denominator and order by the numerator.
- When adding or subtracting fractions, change both to the same denominator and add or subtract the numerators.
- If mixed numbers are involved, deal with the whole numbers separately.

10 Listing events and probability

You should already know

- probabilities are expressed as fractions, decimals or percentages
- all probabilities lie on a scale of 0 to 1 (0 to 100%)
- how to find probability from a set of equally likely outcomes.

Covering all the possibilities

To find the probability of an event happening you need to list all the possible outcomes.

One frequent error with two coins is to think the possible outcomes are '2 heads', '2 tails' or '1 head and 1 tail' and think they are all equally likely.

Try tossing two coins about a hundred times and record what you get. You will find that 1 head and 1 tail occurs about twice as much as each of the other possibilities. This is because '1 head and 1 tail' can be a tail and a head or a head and a tail.

There are two ways of showing all the outcomes.

Table of outcomes

First coin	Second coin
head	head
head	tail
tail	head
tail	tail

This clearly shows there are four outcomes.

Showing outcomes on a grid (a possibility space)

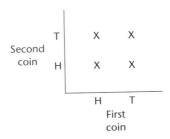

The advantages of this method are:

- it is a very quick way of showing the outcomes when there are very many possibilities, for example it is a very good way to show all 36 outcomes for throwing two dice
- if the outcomes are numerical, the crosses can be replaced by the numerical result. For example, if you are interested in the total score when you throw two dice, you can replace the X for (3, 4) by 7 and the X for (5, 6) by 11 etc.

A disadvantage is that it cannot be extended for a third or subsequent trial. It is difficult to draw a three-dimensional graph but it is easy to add a third column to a table.

When all the possibilities are listed you can find the probability from.

$$\text{Probability of an event} = \frac{\text{Number of outcomes which give the event}}{\text{Total number of outcomes}}$$

In the throwing of two coins listed above, the probabilities are:

Probability of 2 heads = $\frac{1}{4}$

Probability of 1 head and 1 tail = $\frac{2}{4} = \frac{1}{2}$

Probability of 2 tails = $\frac{1}{4}$.

EXAMPLE 1

Fatima throws an ordinary die and tosses a coin. Show the possible outcomes in a table and find the probability that

a) Fatima scores a head and a 6

b) Fatima scores a tail and an odd number.

a) P(head and 6) = $\frac{1}{12}$

b) P(tail and odd number) = P[(T, 1) or (T, 3) or (T, 5)] = $\frac{3}{12}$ = $\frac{1}{4}$

This question could also have been answered using a grid.

```
      T │ X  X  X  X  X  X
Coin    │
      H │ X  X  X  X  X  X
        └─────────────────
          1  2  3  4  5  6
                Die
```

Choose the outcomes logically, changing the number on the die first.

Die	Coin
1	head
2	head
3	head
4	head
5	head
6	head
1	tail
2	tail
3	tail
4	tail
5	tail
6	tail

Therefore there are 12 possible outcomes.

EXAMPLE 2

Gareth tosses two dice. Show the outcomes on a grid (possibility space) and use the grid to find the probability that

a) Gareth scores a double

b) Gareth scores a total of 11.

From the grid, there are 36 possible outcomes.

a) P(double) = $\frac{6}{36}$

$= \frac{1}{6}$

b) P(score of 11) = $\frac{2}{36}$

$= \frac{1}{18}$.

The grid could also have been drawn with the total scores replacing the Xs.

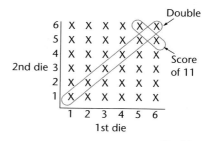

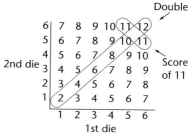

Chapter 10 *Listing events and probability*

EXAMPLE 3

Rachel is selecting a main course and a sweet from this menu.

```
              MENU

   Main Course              Sweet
   Sausage & Chips        Apple Pie
     Ham Salad           Fruit Salad
   Vegetable Lasagne
```

Main course	Sweet
Sausage and Chips	Apple Pie
Sausage and Chips	Fruit Salad
Ham Salad	Apple Pie
Ham Salad	Fruit Salad
Vegetable Lasagne	Apple Pie
Vegetable Lasagne	Fruit Salad

List the possible outcomes in a table. If she is equally likely to select any of the choices, what is the probability that she selects Vegetable Lasagne and Apple Pie?

Therefore there are six possible outcomes.

P(Vegetable Lasagne and Apple Pie) = $\frac{1}{6}$.

EXERCISE 10.1A

1 Jamil has brown socks, red socks and green socks in his drawer. He selects two socks at random from the draw. Copy the table and complete it to show the possible outcomes.

First sock	Second sock

2 In tennis one player must win, a draw is not possible. Alex plays three games of tennis against Meiling. Copy the table and complete it to show the possible winner of each game. The first entry has been done for you.

First game	Second game	Third game
Alex	Alex	Alex

3 Mr and Mrs Green plan to have three children. Draw a table to show the possible sexes for the first, second and third child. Assuming that all the outcomes are equally likely, what is the probability that Mr and Mrs Green will have

a) all girls   **b)** two boys and a girl?

Exercise 10.1A cont'd

4   The picture shows a fair spinner. Claire spins the spinner twice and records the total score.

Draw a grid, with each of the axes marked 1–5, to show the possible outcomes. Find the probability that Claire scored **a)** 10   **b**) 5

5   There are four suits in a set of playing cards: hearts (H), spades (S), diamonds (D) and clubs (C). There are equal numbers of each. Hearts and diamonds are red, spades and clubs are black.

David chooses a card from a set of playing cards. He records the suit, replaces the card and then chooses another. Draw a grid with the axes labelled, H, S, D and C to show the possible outcomes for his two cards.

What is the probability that David chooses:

a) two spades   **b)** two red cards?

6   A spinner at a fair has numbers 2, 4, 6, 8, 10 as shown. Tony has two spins.

a) Draw a grid with 2, 4, 6, 8, 10 on each axis to show the possible outcomes.

b) How many outcomes are there?

c) What is the probability that Tony gets the same number twice?

7   At an Away Weekend students have to do an activity in the morning and an activity in the afternoon.

The morning activities are Trampoline, Table Tennis or Badminton.

The afternoon activities are Canoeing or Climbing.

a) Complete this table to show the possible choices.

b) If all activities are given out at random, what is the probability that John plays badminton and goes canoeing?

Morning	Afternoon

Exercise 10.1A cont'd

8 Soraya chooses a card from a pack of playing cards and records its suit. She then throws a fair die. Draw a grid showing the possible outcomes. What is the probability that Soraya scores

   a) a club and a 5
   b) a red card and a 6
   c) a heart and an even number?

9 Fiona chooses a playing card, records its suit and tosses a coin. Draw a grid to show the possible outcomes. Find the probability that Fiona scores

   a) a club and a head
   b) a red card and a tail.

10 Lisa is choosing from this menu. Make a table to show her possible choices. You could use the initial letters e.g. OB to save you writing the names out in full. Assuming she is equally likely to choose any of the items, find the probability that she chooses Onion Bhaji and Lamb Madras.

> STARTER AND MAIN COURSE £6
>
> MENU
>
> **STARTERS**   **MAIN COURSE**
>
> Onion Bhaji   Chicken Tikka
> Sheek Kebab   Lamb Madras
>   Vegetable Bhuna

EXERCISE 10.1B

1 Brian has a bag of coins. It contains some 1p, some 2p and some 10p. He picks out two coins one after another.

Complete the table to show the possible outcomes. The first line has been filled in. (you may need more lines)

First coin	second coin
1p	2p

2 Roddy tossed a coin and threw a normal dice.

   a) On a grid list the possible outcomes.
   b) Find the probability that he gets
      i) a head and a six
      ii) an odd number and a tail.

Exercise 10.1B cont'd

3 Robbie tosses three coins together. Copy the table and complete it to show all the possible outcomes. The first has been done for you.

First coin	Second coin	Third coin
head	head	head

Find the probability that Robbie tosses

a) two heads and a tail

b) at least one head.

4 Anne plays a game. She tosses a coin to see whether to pick up a number or a letter card at random. If she tosses a head she picks a number card: 1, 2, or 3. If she tosses a tail she picks a letter card: A, B or C. Copy this table and complete it to show her possible outcomes.

Coin	Card

What is the probability that she picks

a) card C

b) a card with an odd number on it?

5 In a game, Bobbie spins both of the spinners. Make a table showing all the possible outcomes. What is the probability of getting a B and a 3?

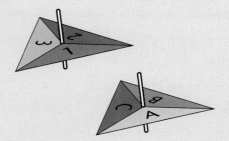

6 Draw a grid showing all the outcomes when two dice are thrown. What is the probability of scoring

a) a double six   **b)** a total score of 9?

7 In the game of Monopoly, you throw two dice and your score is the total. Use the grid you drew for question 6 to answer these questions.

a) To buy Park Lane, Hamish needs to score 11. What is the probability that Hamish can buy Park Lane after his next go?

b) To get out of Jail, Sylvia needs to throw a double . What is the probability that Sylvia gets out of jail on her next go?

c) If Sanjay scores 7, he lands on Regent Street. What is the probability that Sanjay does not land on Regent Street?

Exercise 10.1B cont'd

8 Mr Ahmed is choosing his new company car. He can choose a Rover or a Peugeot. He can choose red, blue or black. Make a table to show all the possible outcomes.

If he chooses completely randomly, what is the probability that he will choose:

a) a blue Peugeot   **b)** a black car?

9 Nicola and John are doing their Maths coursework. They can choose 'Billiard tables', 'Stacking cans' or 'Number trees'.

a) Draw a grid to show all the possible choices of the pair of them.
b) If they choose at random, what is the probability that they both choose the same task?

10 Gary and Salma are going out on Saturday night. They can go bowling, to the cinema or to the fair. After that they can either go for a meal or go dancing. Since they cannot agree what to do they decide to choose randomly.

Make a table to show all the possible choices. Find the probability that they go bowling and then on for a meal.

Key ideas

● **Events can be shown using**

Tables	Advantage	Easy to read for probabilities of equally likely events.
	Disadvantage	Lengthy, cannot be used if events are not equally likely.
Grids	Advantage	Easy to read for probabilities of equally likely events.
	Disadvantages	Lengthy, cannot be used if events are not equally likely.
		Cannot be used for more than two successive events.

● **Probability of an event** = $\dfrac{\text{Number of outcomes which give the event}}{\text{Total number of outcomes}}$

11 Classifying quadrilaterals

You should already know

- a quadrilateral has 4 sides
- the meaning of the words 'parallel', 'adjacent' and 'bisect'
- how to draw quadrilaterals.

Different quadrilaterals

ACTIVITY 1

Draw as many different types of quadrilaterals as you know and label them with their names.

Here are seven different types drawn.

a)

b)

c)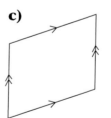

d)

e)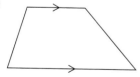

f)

g)

The names and geometrical properties are given in the table below.

| | Name | Angles | Sides | | Diagonals |
			Lengths	Parallel	
a)	Square	All four are 90°	All same length	Opposite sides parallel	Equal length Bisect at 90°
b)	Rectangle	All four are 90°	Opposite sides are the same length	Opposite sides are parallel	Equal length Bisect but not at 90°
c)	Parallelogram	Opposite angles are equal	Opposite sides are the same length	Opposite sides are parallel	Not equal length. Bisect but not at 90°
d)	Rhombus	Opposite angles are equal	All same length	Opposite sides are parallel	Not equal length. Bisect at 90°
e)	Trapezium	Can be all different	Can be all different	One pair of sides parallel	Nothing special.
f)	Kite	One pair of opposite angles equal.	Two pairs of adjacent sides equal.	None parallel	Only one of the diagonals is bisected by the other and they cross at 90°
g)	Isosceles trapezium	Two pairs of adjacent angles equal	One pair of opposite sides equal	One pair of opposite sides parallel	Equal length. Do not bisect or cross at 90°

Many people think that all trapezium are isosceles but any shape that has one pair of opposite sides parallel is a trapezium.

EXAMPLE 1

Which quadrilaterals have
a) both pairs of opposite sides equal?
b) just one pair of opposite angles equal?
a) Square, Rectangle, Parallelogram, Rhombus.
b) Kite.

Chapter 11 *Classifing quadrilaterals*

EXERCISE 11.1A

1 Which quadrilaterals have all angles the same?

2 Which quadrilaterals have opposite sides the same, but not all four sides the same?

3 A rectangle can be described as a special type of parallelogram. What extra thing is true about a rectangle?

4 a) Draw a square and clearly mark any angles that are the same and any sides that are the same.

  b) Draw in the diagonals and check what is true about them.

5 a) Draw a kite and clearly mark any angles that are the same and any sides that are the same.

  b) Draw in the diagonals and check what is true about them.

6 A quadrilateral has its four angles as 80°, 80°, 100° and 100° as you go round the quadrilateral.

  a) Sketch the quadrilateral.

  b) What special type of quadrilateral is this?

7 a) Draw a trapezium that is not isosceles. What can you say about the angles?

8 Which quadrilaterals have diagonals that cross at 90°?

9 Name these shapes.

  a)
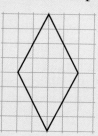

  b)
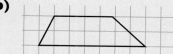

  c)

  d)
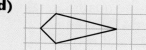

10 Draw a large rectangle on squared paper. Draw lines to split it into as many of the different quadrilaterals as you can (you may be left with some triangles as well). Write the names in all the shapes.

EXERCISE 11.1B

1 Which quadrilaterals have all the sides the same?

2 Which quadrilaterals have opposite angles the same, but not 90°?

3 A rhombus can be described as a special type of parallelogram. What extra thing is true about a rhombus?

4 a) Draw a rectangle and clearly mark any angles that are the same and any sides that are the same.
  b) Draw in the diagonals and check what is true about them.

5 A quadrilateral has the four angles 80°, 100°, 80° and 100° as you go round the quadrilateral. The sides are not all the same length.
  a) Sketch the quadrilateral.
  b) What special type of quadrilateral is this?

6 a) Draw a parallelogram and mark clearly any sides that are the same or parallel and any angles that are the same.
  b) Draw in the diagonals and check what is true about them.

7 A quadrilateral has the diagonals the same length. What types of quadrilateral could it be?

8 List all the quadrilaterals that have both pairs of opposite sides parallel and equal.

9 Name these shapes.

a)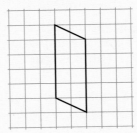

b)

c)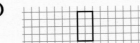

d)

10 Look around at school and at home and find as many of the different quadrilaterals as you can. List them and where they are.

Key ideas

- Quadrilaterals have four sides.
- Many quadrilaterals have special names. Learn what is special about the seven different ones named in this chapter.

12 Simplifying algebra

You should already know

- that letters can be used for numbers
- how to add and subtract numbers.

ACTIVITY 1

Work these out.

1 $3 + 4 + 5 - 2$ **2** $2 - 5 - 4 + 8$

3 $9 - 1 - 2 - 7 + 4$ **4** $6 - 8$

5 $9 - 4 - 3 + 2 - 6$

Remember it is easier to collect all the + numbers and then the − numbers and find the difference. This is also true of letters.

EXAMPLE 1

Work out   **a)** $2a - 3a + 4a$ **b)** $5a - 2a + 4a - 6a$ **c)** $5b - 4b - 8b + 2b.$

Answers   **a)** $2a + 4a - 3a$

$= 6a - 3a$

$= 3a$

Collect all the + and − separately
then add up each and find the difference.

b) $5a + 4a - 2a - 6a$

$= 9a - 8a$

$= a$

c) $5b + 2b - 4b - 8b$

$= 7b - 12b$

$= {}^-5b$

Making it simple

Here is the formula for the perimeter P of this rectangle.

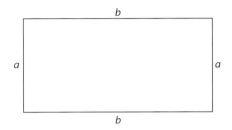

> **Exam tip**
>
> You can only simplify when adding (or subtracting) the same letter(s), for example, $a + 2a = 3a$, $4ab - 3ab = ab$ and $ab + bc = ab + bc$

$P = a + a + b + b$

$a + a$ is twice as big as a.

So $a + a = 2 \times a$, which is usually written as $2a$, meaning '2 times a'.

So $P = 2a + 2b$.

Here is the formula for the area of the rectangle.

$A = a \times b$.

The shortest way of writing $a \times b$ is ab.

Here is the formula for the area of this square.

> **Exam tip**
>
> $1a$ or $1 \times a$ is always written as just a.
> Some people confuse $2a$ and a^2.
> $2a = 2 \times a = a + a$, while $a^2 = a \times a$.
> Writing the multiplication sign in helps:
> $2a \times 3b = 2 \times a \times 3 \times b = 2 \times 3 \times a \times b = 6ab$.

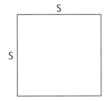

$A = s \times s$.

The short way of writing $s \times s$ is s^2 (said as 's to the power of 2' or 's squared').

EXAMPLE 2

$3a + 2b - 2a + 5b = 3a - 2a + 2b + 5b$
$\qquad\qquad\qquad\quad = a + 7b$

EXAMPLE 3

$p \times q \times p^2 \times q = p \times p^2 \times q \times q$
$\qquad\qquad\qquad\quad = p^3 \times q^2$
$\qquad\qquad\qquad\quad = p^3 q^2$ said as 'p cubed q squared'

EXERCISE 12.1A

Simplify these.

1 $x + x + x + x + x$
2 $y + y + y + z + z$
3 $a \times a \times a$
4 $a \times a + b \times b$
5 $p + p + q + p + q + q + p$
6 $a + 2b + 2a + b$
7 $3m + 2n + n + m$
8 $5x - 3x + 2y - y$
9 $6p + 2q - 5q - 3p$
10 $a^2 + 2a - 3a - 6$
11 $a + a + a \times a$
12 $2b + 3b$
13 $2b \times 3b$

14 A square has sides $2a$ long. Find and simplify an expression for its perimeter.

15 A rectangle has width $2b$ and length $3a$. Find and simplify an expression for its area.

EXERCISE 12.1B

Simplify these.

1 $x + x + y + y$
2 $a + b + a + b + a$
3 $2p + 3q + 2q + 5p$
4 $2m + 3n - m - n$
5 $x^2 + 3x - 5x - 15$
6 $3p \times 4q$
7 $5x \times 2y + x \times 3y$
8 $xy^2 \times xy$
9 $3x^2 + 2y^2 - 5x^2$
10 $2a \times 3b$
11 $2a \times b + a \times 3b$
12 $2a - 3b + 4a + b$
13 $a + 6b - 5a + 2b$

14 A quadrilateral has sides $2a$, $3b$, $4a$ and $6b$. Find and simplify an expression for its perimeter.

15 David bought five bottles of tizer at $2a$ pence each and three packets of chocolate at a pence each. Find and simplify an expression for the total amount he spent.

Collecting like terms and simplifying expressions

Remember that $a + a + a = 3a$

$$4a + 3b + b - a = 3a + 4b$$
$$a \times b = ab$$
$$a \times a \times a = a^3$$
$$p^2 \times p = p \times p \times p = p^3$$

Complicated terms such as ab^2, a^2b and a^3 cannot be collected together unless they are exactly the same type.

EXAMPLE 3

Simplify these expressions by collecting together the like terms.

a) $2x^2 - 3xy + 2yx + 3y^2$
b) $3a^2 + 4ab - 2a^2 - 3b^2 - 2ab$
c) $3 + 5a - 2b + 2 + 8a - 7b$.

a) $2x^2 - 3xy + 2yx + 3y^2 = 2x^2 - xy + 3y^2$
The two middle terms are like terms because xy is the same as yx.

b) $3a^2 + 4ab - 2a^2 - 3b^2 - 2ab = a^2 + 2ab - 3b^2$
Here the a^2 terms, the ab terms and the single b^2 term can be collected but they cannot be combined.

c) $3 + 5a - 2b + 2 + 8a - 7b = 5 + 13a - 9b$
Here there are three different types which can be collected separately but not together.

Exam tip

Errors are often made by trying to go too far, for example $2a + 3b$ cannot be simplified any further.
A further common error is to work out $4a^2 - a^2$ as 4. The answer is $3a^2$.
Remember that ab is the same as ba.

EXERCISE 12.2A

Simplify where possible.

1 $2a + 3b - 2b + 3a$
2 $4ab - 3ac + 2ab - ac$
3 $a^2 + 3b^2 - 2a^2 - b^2$
4 $2x^2 - 3xy - xy + y^2$
5 $4b^2 + 3a^2 - 2b^2 - 4a^2$
6 $5a^3 + 4a^2 + 3a$
7 $a^3 + 3a^2 + 2a^3 + 4a^2$
8 $9abc + 4cab - 5bca + 6cba$
9 $3x^3 - 2x^2 + 4x^2 - 3x^3$
10 $a + 3b - b + 2a - 3a - 2b$

EXERCISE 12.2B

Simplify where possible.

1 $3a - 4b + 2a - 2b$
2 $9a^2 - 3ab + 5ab - 6b^2$
3 $4ab + 2bc - 3ba - bc$
4 $2p^2 - 3pq + 4pq - 5p^2$
5 $3ab + 2ac + ad$
6 $9ab - 2bc + 3bc - 7ab$
7 $a^3 + 3a^3 - 6a^3$
8 $3ab^2 - 4ba^2 + 7a^2b$
9 $8a^3 - 4a^2 + 5a^3 - 2a^2$
10 $abc + cab - 3abc + 2bac$

Key ideas

- Terms can only be added or subtracted if they are of exactly the same type.
- ab and $2ab$ can be added but a^2 and $2a$ cannot.

C1 Revision exercise

1 Fill in the blanks in these equivalent fractions.

a) $\frac{1}{4} = \frac{2}{\square} = \frac{\square}{16} = \frac{6}{\square}$

b) $\frac{3}{5} = \frac{6}{\square} = \frac{\square}{25} = \frac{18}{\square}$.

2 Write these fractions in their lowest terms.

a) $\frac{3}{9}$ **b)** $\frac{15}{35}$ **c)** $\frac{8}{12}$ **d)** $\frac{18}{54}$.

3 Write these fractions in order, smallest first.

$\frac{7}{20}$ $\frac{3}{4}$ $\frac{17}{20}$ $\frac{4}{5}$

4 Work these out.

a) $\frac{1}{2} + \frac{1}{4}$ **b)** $\frac{5}{8} + \frac{3}{4}$

c) $\frac{5}{8} - \frac{1}{4}$ **d)** $1\frac{5}{8} + 2\frac{1}{4}$

e) $2\frac{5}{8} - \frac{3}{5}$.

5

> **MENU**
>
> | First course | Second course |
> |---|---|
> | Sausage & chips | Apple Pie |
> | Ham Salad | Fruit Salad |
> | Vegetable Lasagne | Ginger Sponge |

This is the menu at Fred's cafe. Julian is going to have a two-course meal.

a) Make a table with two columns and list all the possible choices he could make.

b) If all the choices are equally likely what is the probability he

   (i) chooses Ham Salad and Ginger Sponge,

   (ii) does not choose Sausage and Chips?

6 Salma throws two dice and records the result of multiplying the two scores together.

Draw a grid to show all the possible outcomes. Find the probability that Salma's results is

a) 36   **b)** 12   **c)** 4.

7 What is the name of a quadrilateral that has

a) all angles the same, and just the opposite sides equal

b) all the sides the same, and just the opposite angles equal?

8 Draw an isosceles trapezium. Mark any sides that are parallel, any sides that are the same and any angles that are the same.

9 A quadrilateral has diagonals that bisect. What types of quadrilateral could it be?

10 Write these as simply as possible.

a) $a + a + a$

b) $p + q + p + q$

c) $3a + 2b - a - 2b$

d) $3a \times 4a$

e) $pq + 2pq$

f) $ab^2 \times a^2b$.

11 Simplify these expressions.

a) $4x - 2y + 3y - 2x$

b) $a^2b + 2ab + 3a^2b - ab$

c) $8y \times 3z$

d) $pq \times p^2q$

e) $3x^2 \times 2xy$.

12 Simplify these by collecting like terms.

a) $3a + 4b + 2a - 4b$

b) $5ab^2 - 2a^2b + 3a^2b - 4ab^2$

c) $2ab + 3ac - 4ad + 2ab + 4ad - ac$

d) $x^2 - 2xy + 3yx + 3x^2$.

13 Plans and elevations

Plans and elevations

Any one view of a three-dimensional object will not show all its features. To get the full picture you need three different views from three perpendicular directions.

The three views are called   the PLAN

the FRONT ELEVATION

the SIDE ELEVATION.

EXAMPLE 1

Sketch the plan and elevations of the solid below.

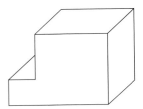

Example 1 cont'd

The PLAN view of the solid shape is the view of the surfaces that you get if you look at the solid directly from above. This is sometimes referred to as the 'birds-eye' view.

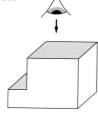

plan

Exam tip

Notice that the only indication of the step in the solid is a line on the Plan. This view will not tell you how deep the step in the solid is.

The FRONT ELEVATION, or front view, of the solid shape is the view of the surfaces that you get if you look at the solid directly from the front.

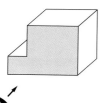

front elevation

Exam tip

Notice that this view will not tell you how wide the solid is.

The SIDE ELEVATION, or side view, of the solid shape is the view of the surfaces that you get if you look at the solid directly from the side.

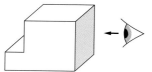

side elevation

Exam tip

The step in the solid is hidden from view from this direction. To show it is there, but cannot be seen, we use a dotted line. Dotted lines are always used to indicate hidden features of the solid.
The side elevation can be viewed from either side of the solid. From the opposite side, the view would be the same but with a solid line instead of a dotted line: the step in the solid *can* be seen from this direction.

EXAMPLE 2

Sketch the plan, P, the front elevation, F, and the side elevation, S, of a ring.

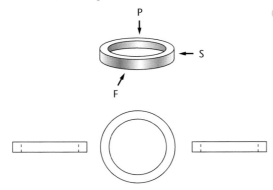

Front elevation   Plan   Side elevation

Exam tip

Notice that the front and side elevations are the same. The two dotted lines indicate the hidden edge of the inside of the ring. The curved part of the ring appears only as a rectangle in the front and the side elevations.

EXAMPLE 3

Draw accurately the plan, P, the front elevation, F, and the side elevation, S, of the small wedge shown in the diagram below.

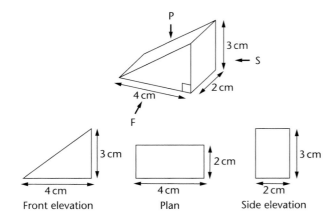

EXERCISE 13.1A

Sketch the plan and elevations of the following solid shapes.
The arrows indicate the direction of the plan, P, the front elevation, F, and the side elevation, S.

1

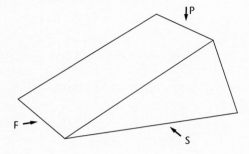

2

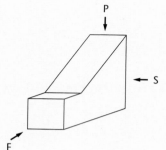

3

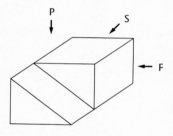

4

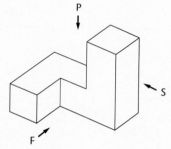

5

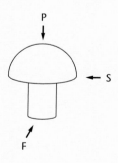

6

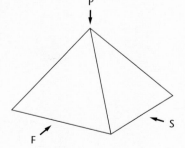

Exercise 13.1A cont'd

7

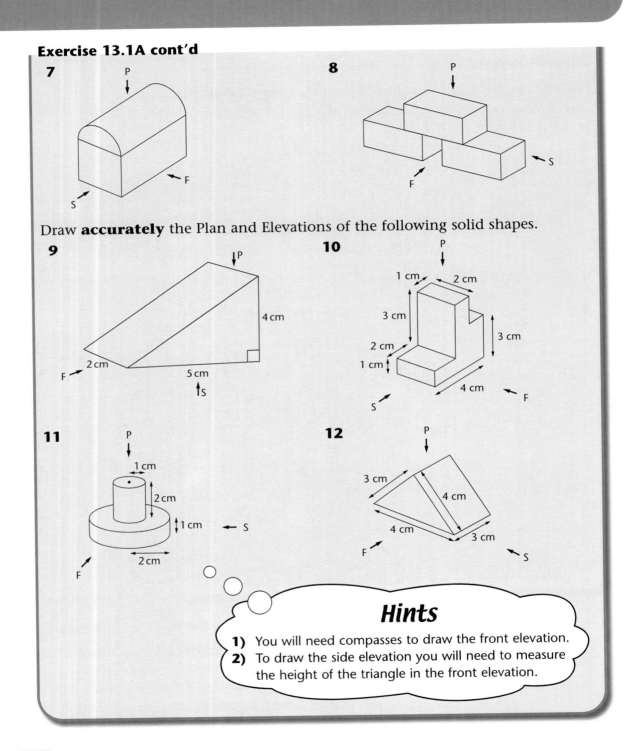

8

Draw **accurately** the Plan and Elevations of the following solid shapes.

9

10

11

12

Hints

1) You will need compasses to draw the front elevation.
2) To draw the side elevation you will need to measure the height of the triangle in the front elevation.

EXERCISE 13.1B

Sketch the plan and elevations of the following solid shapes.

The arrows indicate the direction of the plan, P, the front elevation, F, and the side elevation, S.

1

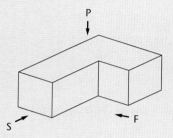

2

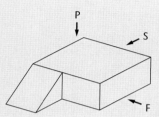

3

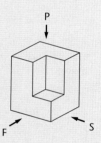

4

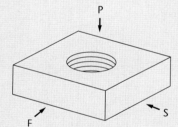

5

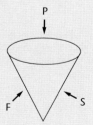

6

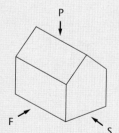

Exercise 13.1B cont'd

7

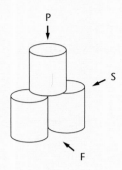

Draw **accurately** the plan and elevations of the following solid shapes.

8

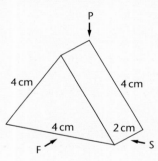

9

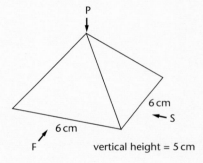

vertical height = 5 cm

10

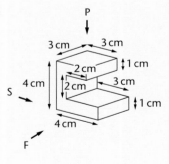

11

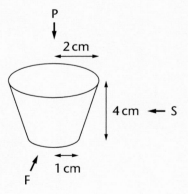

Exercise 13.1B cont'd

12

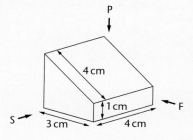

Key idea

● Accurate drawings must be accurate, use a sharp pencil and check all measurements.

14 Interpreting statistical data and measures

You should already know

- how to find the median, mean, mode and range of a set of data
- how to draw and measure angles.

Grouping data

When you have to analyse a lot of data it is sometimes helpful to group or sort the data into bands, or intervals, of equal width. This makes it easier to find the frequencies, especially if you draw up a tally chart,

ACTIVITY 1

Collect information from pupils in your class on their spending money. Group into suitable amounts and make a tally chart.

EXAMPLE 1

Here are the prices of irons from the catalogue of an electrical superstore.

£9·60	£12·95	£13·90	£13·95	£14·25
£16·75	£16·90	£17·75	£17·90	£19·50
£19·50	£21·75	£22·40	£22·40	£24·50
£24·90	£26·00	£26·75	£26·75	£27·50
£29·50	£29·50	£29·50	£32·25	£34·25
£34·25	£34·50	£35·25	£35·75	£38·75
£39·00	£39·50	£47·00	£49·50	

The prices can be grouped into bands that are £5 wide.

Price (£)	Tally	Frequency
5·00–9·99	I	1
10·00–14·99	IIII	4
15·00–19·99	IIII I	6
20·00–24·99	IIII	5
25·00–29·99	IIII II	7
30·00–34·99	IIII	4
35·00–39·99	IIII	5
40·00–44·99		0
45·00–49·99	II	2
Total		34

Exam tip

Although filling in a tally chart takes a little time, it is well worth the effort, as you are less likely to miss out values as you work through a set of data. If you make a mark for every entry, drawing every fifth stroke through the previous four, you can count up the values in fives, which should be quick and easy.

Exam tip

Unless you are told what size bands or intervals to use, you should normally try to divide the values up into no more than about eight equally-sized bands.

EXAMPLE 2

The table shows the marks gained by the 26 children in class 6, in test A of the Key Stage 2 SATs.

28	36	6	17	24	21	24	12	19
27	30	13	19	9	18	35	12	26
9	27	13	35	15	8	25	14	

Their teacher groups the marks into intervals of 10.

Mark	Tally	Frequency
0–9	IIII	4
10–19	HH HH	10
20–29	HH III	8
30–39	IIII	4
	Total	26

The frequencies of all the sets of data can be plotted in a diagram like a bar chart, which is called a frequency chart of a frequency diagram.

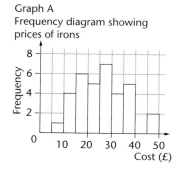

Graph A
Frequency diagram showing prices of irons

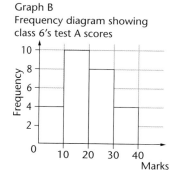

Graph B
Frequency diagram showing class 6's test A scores

EXERCISE 14.1A

Remember $20 \leqslant x < 30$ includes 20 but not 30.

1 These are the heights of 20 girls in Form 5B in centimetres.

| 155 | 145 | 153 | 156 | 162 | 161 | 147 | 142 | 143 | 134 |
| 137 | 139 | 143 | 140 | 157 | 151 | 154 | 158 | 138 | 144 |

Using groups 130 to 134, 135 to 139 etc., make a tally chart of these heights.

2 The times estimated, by 50 ten-year-old children, for 1 minute are given in this table.

Time (seconds)	$20 \leqslant x < 30$	$30 \leqslant x < 40$	$40 \leqslant x < 50$	$50 \leqslant x < 60$	$60 \leqslant x < 70$	$70 \leqslant x < 80$
Frequency	0	8	14	20	7	1

Draw the frequency diagram for these times.

3 These are the ages of 30 members of Heartbeat health club.

25	24	28	29	34	35	38	39	65	63
27	35	48	46	59	40	67	61	52	54
58	42	28	21	47	36	25	54	59	61

a) Using groups $20 \leqslant x < 30$ etc., make a tally chart (frequency table) of these ages.

b) Draw the frequency diagram of these ages.

4 These are the marks of 30 pupils in Form 4A in a test.

32	42	23	37	29	12	37	6	5	37
17	18	31	29	27	11	21	37	28	37
22	31	23	47	23	12	24	34	41	43

a) Using intervals of $0 \leqslant x < 10$, $10 \leqslant x < 20$ etc., make a tally chart (frequency table) of these marks.

b) Draw the frequency diagram of these marks.

5 The heights of 20 women at a luncheon club were recorded. These are the heights in centimetres.

| 155 | 147 | 158 | 171 | 150 | 159 | 165 | 164 | 172 | 154 |
| 170 | 162 | 153 | 157 | 149 | 145 | 171 | 148 | 173 | 178 |

a) Using intervals $140 \leqslant x < 150$, $150 \leqslant x < 160$ etc., make a frequency table of these heights.

b) Draw a frequency diagram of these heights.

EXERCISE 14.1B

Remember $20 \leqslant x < 30$ includes 20 but not 30.

1 These are the masses of 20 people in kilograms.

67	68	79	85	81	62	67	64	65	63
74	75	78	81	76	65	58	57	61	71

Use groups 50–54, 55–60 etc., to make a tally chart of this data.

2 This table gives the spending money received by 25 students.

Spending money (£)	$0 \leqslant x < 5$	$5 \leqslant x < 10$	$10 \leqslant x < 15$	$15 \leqslant x < 20$	$20 \leqslant x < 25$	$25 \leqslant x < 30$
Frequency	0	5	12	4	3	1

Draw a frequency diagram of this data.

3 These are the times, in seconds, athletes took to run 800 metres.

118	122	125	124	126	128	116	114	115	118
117	132	128	136	121	136	138	130	120	115

a) Using groups $110 \leqslant x < 115$, $116 \leqslant x < 120$ etc., make a tally chart (frequency table) of these times.

b) Draw the frequency diagram of these times.

4 These are the ages of the first 20 people to answer a questionnaire.

44	48	61	28	37	42	57	65	75	44
25	36	71	64	63	58	45	47	51	62

a) Using groups $20 \leqslant x < 30$ etc., make a tally chart (frequency table) of these ages.

b) Draw the frequency diagram of these times.

5 These are the lengths of 22 students longest finger in millimetres.

53	66	75	61	73	62	79	56	71	54	58
75	71	63	68	53	64	69	72	54	53	60

a) Using intervals of 50–54, 55–59 etc., make a frequency table of these lengths.

b) Draw the frequency diagram of these lengths.

Mean, median, mode and range of data

You should already know how to find the mode, which is the most common value in a set of data or numbers in a list of numbers.

When dealing with grouped data you will be expected to find the modal class rather than a single value. This will be the group with the greatest frequency. On a frequency chart it is the group with the highest bar.

If you look back to page 111, in the graph for the prices of irons, you will see that the modal class is £25–29·99 because there are more irons for sale within that price range. The graph for the pupils' marks shows a modal class of 10–19, with 10 pupils.

EXERCISE 14.2A

Look back at your answers to Exercise 14.1A questions 1 to 5 and write down the modal class for each set of data.

EXERCISE 14.2B

Look back at your answers to Exercise 14.1B questions 1 to 5 and write down the modal class for each set of data.

Calculating mean, median, mode and range

You should be able to calculate the mean, median, mode and range for a set of data.

EXAMPLE 3

The label of a matchbox is marked 'Average contents 50 matches'. A survey of ten matchboxes produced the following data.

Box	A	B	C	D	E	F	G	H	I	J
Number of matches	45	51	48	47	46	47	49	50	52	47

Find the range, mean, median and mode of the contents.

Writing the numbers in size order gives:

45   46   47   47   47   48   49   50   51   52

Mode (the number that occurs most often) = 47

Example 3 cont'd

Mean = $\dfrac{45 + 46 + 47 + 47 + 47 + 48 + 49 + 50 + 51 + 52}{10} = \dfrac{482}{10} = 48 \cdot 2$

Median (the middle number, because there are ten numbers in the data the median lies halfway between the middle two numbers) = $\dfrac{47 + 48}{2} = 47 \cdot 5$

Range (the difference between the smallest and the largest values) = $52 - 47 = 7$

If data are presented in a frequency table, you can work out the range, mean, median and mode without needing to work on all the data.

EXAMPLE 4

One day in September, a scientist recorded the highest temperature at 48 coastal towns.

Temperature (°C)	14	15	16	18	19	20	21	22	23	24	25	27
Number of towns (frequency)	1	1	2	2	8	12	10	4	3	2	2	1

- The mode is 20°C because this is the temperature with the greatest frequency.
- The median is midway between the 24th and 25th numbers. You could write out the temperatures: 14, 15, 15, 16, 18, 18, 19, 19, 19 but it is quicker to add up the frequencies until you reach the place where the next frequency added would be greater than the middle value, here 25.

  So adding up the first five frequencies gives: 1 + 1 + 2 + 2 + 8 = 14.

  Adding on the next frequency, 12, would give a total of 26 so the 24th and 25th values must occur in the band where the frequency is 12 and thus the median temperature is 20°C.
- The mean is worked out by multiplying each temperature by its frequency and dividing by the total frequency:

  $\dfrac{\begin{array}{l} 14 \times 1 + 15 \times 1 + 16 \times 2 + 18 \times 2 + 19 \times 8 + 20 \times 12 + \\ 21 \times 10 + 22 \times 4 + 23 \times 3 + 24 \times 2 + 25 \times 2 + 27 \times 1 \end{array}}{48} = 20 \cdot 4°C$
- The range = 27°C − 14°C = 13°C

Exam tip

Remember to put the data in size order when you are finding the median.

Chapter 14 *Interpreting statistical data and measures*

EXERCISE 14.3A

1 For each of these sets of data work out the mean, median and mode.
Give the mean to one decimal place.

a) 2 5 5 6 9 5 4 6 7   **b)** 11 13 15 17 19 14 12 15

c) 7 8 1 9 5 8 8 2 3 4   **d)** 19 15 18 17 16 14 19

e) 234 235 239 221 274 271 266 232

2 For these golf scores work out the mean, median and mode.

Score	2	3	4	5	6	7
Frequency	2	0	4	4	7	1

3 For these marks work out the mean, median and mode.

Mark	1	2	3	4	5	6	7	8	9	10
Frequency	0	1	3	0	6	5	2	4	3	1

4 A football team played 26 matches during a season. The number of goals they scored were as follows.

Number of goals	0	1	2	3	4	5	6
Number of matches (frequency)	3	5	6	7	2	2	1

Find the mean, median, mode and range of the number of goals per match.

5 The team decides to measure the fitness of its players.

One test is to run 800 metres. These are the times, in seconds, that the players recorded.

136·4   186·0   152·6   178·0   157·8   150·8   151·6   162·4

166·0   152·4   167·2   138·2   149·8   176·0   146·4

Calculate the mean, median, mode and range of the times.

EXERCISE 14.3B

1 For each of these sets of data work out the mean, median and mode.
   Give the mean to one decimal place.
   a) 4 7 7 8 11 7 6 8 9   **b)** 21 23 25 27 29 24 22 25
   c) 6 7 0 8 4 7 7 1 2 3   **d)** 9 5 8 7 6 4 9
   e) 534 535 539 521 574 571 566 532

2 These are the hours worked in a week by Jonathan over 20 weeks.
   Work out the mean, median and mode.

Hours worked	36	37	38	39	40	41	42	43	44
Frequency	2	0	6	4	3	1	1	2	1

3 For these marks work out the mean, median and mode.

Mark	1	2	3	4	5	6	7	8	9	10
Frequency	3	2	4	5	7	2	0	1	0	1

4 The table shows the scores of 40 players after the first round of a
   golf tournament.
   70 68 71 67 74 69 69 71 68 70 71 70 72 69 69 68 71 70 70 72
   72 69 68 70 68 69 67 71 69 70 68 67 70 70 73 69 71 67 69 68
   Make a frequency table and calculate the mean, median, mode and range of
   the scores.

5 Thirty pupils estimated a 30cm length. Here are their estimates.
   29·0   30·3   30·9   29·0   26·7   28·3   30·0   28·9   28·6   23·4
   30·0   26·2   27·2   27·5   29·4   30·5   25·0   24·6   26·5   27·5
   20·0   19·4   23·0   20·0   22·0   20·2   23·8   34·0   22·5   20·5
   Calculate the mean, median, mode and range of the estimates.

Comparing data

EXAMPLE 5

The children in a swimming club record the time it takes them to swim one length of the pool, backstroke.
The results are:

	Time (seconds)															
Boys	45	46	48	60	42	53	47	51	54	54	49	48	47	53	48	45
Girls	45	47	47	55	46	53	54	63	48	50	46	51	48	48		

Compare the distributions of the times for boys and girls.

Boys: mean = 49   Girls: mean = 50

     median = 48   median = 48

     mode = 48   mode = 48

     range = 60 − 42 = 18   range = 63 − 45 = 18

For both boys and girls the median, mode and range are the same but the mean for the girls is slightly higher
so you could deduce that the girls are slightly slower than the boys.

EXERCISE 14.4A

1 Brian and Les played a round of golf. Their scores had the following
mean, median, mode and range.

Brian   Mean = 4·9, median = 5, mode = 6, range = 6.

Les   Mean = 5·2, median = 5, mode = 7, range = 4.

The better player gets lower scores. Who was the better player?
Why is he better?

2 In the Dronfield singers, the women have a mean age of 51, a median
age of 47 and a range of 34. The men have a mean age of 48, a median
age of 45 and a range of 28. Compare the ages of the women and the
men.

3 The marks of two groups of students in French had the following
mean, median, mode and range.

A   Mean = 43·6, median = 46, mode = 46, range = 17

B   Mean = 43·8, median = 47, mode = 46, range = 13.

Compare the marks of the two groups.

Exercise 14.4A cont'd

4 The number of goals scored in 20 matches by Devon Rovers is given in this table.

Number of goals	0	1	2	3	4	5
Frequency	5	6	4	3	1	1

a) Work out the mean, median, mode and range for Devon Rovers.

b) In their 20 games, Melton Rangers statistics were Mean = 1·8, mode = 1, median = 2, range = 6. Compare the results of the two teams.

EXERCISE 14.4B

1 In the Dearne cricket eleven, Bostwick and Hodgson are two of the batsmen. In the first 10 matches of the season their statistics were:

Bostwick: Mean 31, median, 28, range 60

Hodgson: Mean 23, median 25 range 42.

a) (i) Who was the better batsman? **(ii)** What makes you pick that person?
b) (i) Who is more consistent? **(ii)** Why do you say that?

2 These are the statistics of the number attending Millhouses youth club.

In 1996, mean = 28·5, median 26 and range 16.

In 2001, mean = 24·2, median = 24 and range 9.

Compare the attendance at the youth club in two years.

3 At Prothero fabric factory the salaries have a mean of £25 000, median of £20 000 and a range of £21 000.

At Jaline fabric factory the salaries have a mean of £23 250, median of £20 000 and a range of £10 000. Compare the salaries at the two factories.

4 a) At the New Friends group, the ages of the women are:

65   78   81   89   92   75   79   82   91   72
89   91   81   83   78.

Work out the mean, median and range of their ages.

b) At the same group the ages of the men are:

75   78   89   84   81   72   74   78

Work out the mean, median and range of their ages.

c) Compare the ages of the men and the women.

Key ideas

In a set of data:

- The mode is the number or value that occurs most often.
- The mean is the usual average and is found by adding up all the values and dividing by the number of values.
- The median is the middle value or, if there is an even number of values, it is halfway between the middle two values.
- The range is the difference between the smallest and the largest values.

15 Translations

You should already know

- how to plot points in all four quadrants
- what congruent means.

When a shape is translated all its points move the same distance in the same direction. The image and the original shape are congruent. For example, this shape ▷ could translate to this ▷.

ACTIVITY 1

This is a pattern made up by translations. Copy it and add 4 more shapes.

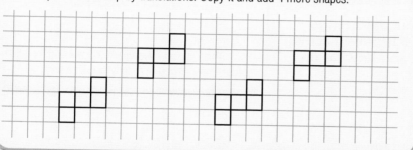

ACTIVITY 2

Draw some more patterns made with translations.

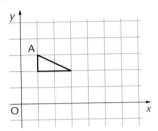

EXAMPLE 1

Translate the shape four squares to the right and two squares down.

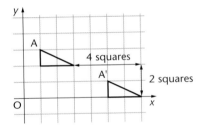

Point A has moved 4 along and 2 down and the other corners have moved in exactly the same way.

Exam tip

A common error is to move the wrong number of squares. Check by taking one point on the original shape and carefully counting to the corresponding point on the image (the translated shape).

Describing translations

To describe a translation you need to say how far the shape has moved across to the left or the right and how far up or down.

EXAMPLE 2

Describe fully the transformation of T to T'.

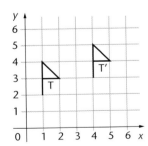

The image is the same way round and the same way up as the original shape, so it is a translation.

It has moved 3 squares to the right and 1 up. Check by counting squares.

This is a translation of three squares to the right and one up.

EXERCISE 15.1A

Make a copy of this diagram and answer questions 1 to 5.

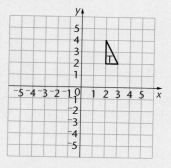

1 Translate the triangle T 2 squares to the right and 1 up. Label it A.

2 Translate the triangle T 1 square to the right and 3 down. Label it B.

3 Translate the triangle T 4 squares to the left and 1 up. Label it C.

4 Translate the triangle T 6 squares to the left and 6 down. Label it D.

5 Translate the triangle T 2 squares to the left and 5 down. Label it E.

Use this diagram for the next 4 questions.

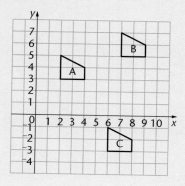

6 Describe the translation that maps shape A onto shape B.

7 Describe the translation that maps shape A onto shape C.

8 Describe the translation that maps shape C onto shape B.

9 Describe the translation that maps shape C onto shape A.

10

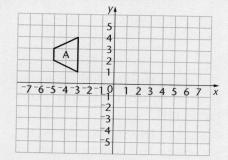

Copy the diagram onto squared paper.

a) Translate shape A 2 to the left and 6 down, label it B.

b) Translate shape B 9 to the right and 5 up, label it C.

c) Translate shape C 2 to the right and 4 down, label it D.

d) What must you translate D by to get back to A?

EXERCISE 15.1B

Make a copy of this diagram and answer questions 1 to 5.

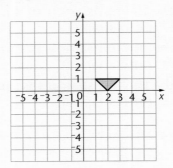

Use this diagram for the next 4 questions.

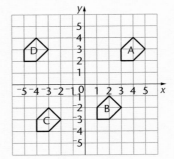

1 Translate the shaded triangle 2 squares to the right and 4 up. Label it A.

2 Translate the shaded triangle 1 square to the right and 5 down. Label it B.

3 Translate the shaded triangle 4 squares to the left and 3 down. Label it C.

4 Translate the shaded triangle 6 squares to the left and 3 up. Label it D.

5 Translate the shaded triangle 3 squares to the left and 4 down. Label it E.

6 Describe the translation that maps shape A onto shape B.

7 Describe the translation that maps shape A onto shape C.

8 Describe the translation that maps shape A onto shape D.

9 Describe the translation that maps shape D onto shape C.

Exercise 15.1B cont'd

10

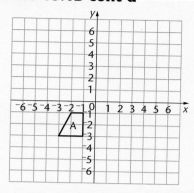

Copy the diagram onto squared paper.

a) Translate shape A 3 to the left and 6 up, label it B.

b) Translate shape B 7 to the right and 8 down, label it C.

c) Translate shape C 1 to the right and 5 up, label it D.

d) Translate shape D 5 to the left and 3 down. What do you notice?

Key ideas

- A translation moves every point on the shape the same distance, in the same direction.
- A translation is defined by movements to the left or right and up or down.

16 Fractions, decimals and percentages

You should already know

- fraction, decimal and percentage notation
- multiplication and division of decimals by 100
- how to round an amount of money to the nearest penny
- how to multiply and divide by simple numbers without a calculator
- how to find 10%, 20% etc. without a calculator.

ACTIVITY 1

Write these percentages as decimals.

a) 50% **b)** 20% **c)** 25%
d) 75% **e)** 100%

Equivalence of fractions, decimals and percentages

You know already that $\frac{1}{2}$, 0·5 and 50% all mean the same.
You can say: 'Half the children in the school are girls'
or 		'0·5 of the children in the school are girls'
or 		'50% of the children in the school are girls'
You may also realise that $\frac{1}{4}$, 0·25 and 25% all mean the same. In this section you will look at other fractions and their decimal and percentage equivalents. You will also see how to convert from one form to another.

Fraction to decimal

Looking back to the case of $\frac{1}{2}$ think of the fraction line between the 1 and 2 as a division (÷) sign. This means that $\frac{1}{2}$ is the same as 1 ÷ 2 which is equal to 0·5.
So the decimal equivalent of $\frac{1}{2}$ is 0·5.

Decimal to percentage

Look again at the two cases above.

0·5 is the same as 50% because 'per cent' means 'out of 100' and 50 out of 100 = $\frac{1}{2}$ = 0·5.

0·25 is the same as 25% because 25 out of 100 = $\frac{1}{4}$ = 0·25.

Now 0·5 × 100 = 50 and 0·25 × 100 = 25.

So to change a decimal to a percentage, simply multiply the decimal by 100.

EXAMPLE 1

Convert $\frac{3}{5}$ to   **a)** a decimal   **b)** a percentage.

a) $\frac{3}{5}$ = 3 ÷ 5 = 0·6 (This can be done on a calculator, or as $5\overline{)3·0}$ with 0·6 above)

b) 0·6 × 100 = 60%

NB It is possible to go straight from the fraction to the percentage using multiplication of fractions.

$\frac{3}{5}$ × 100 = $\frac{3}{5}$ × $\frac{100}{1}$ = $\frac{300}{5}$ = 60%

EXERCISE 16.1A

1 Change the following fractions to decimals:

   a) $\frac{8}{10}$ **b)** $\frac{4}{5}$ **c)** $\frac{1}{4}$

   d) $\frac{7}{20}$ **e)** $\frac{17}{50}$

2 Change the decimals you found in question 1 to percentages.

3 Change the following fractions to percentages:

   a) $\frac{3}{10}$ **b)** $\frac{2}{5}$ **c)** $\frac{3}{4}$

   d) $\frac{19}{20}$ **e)** $\frac{9}{50}$

4 Change the following percentages to decimals:

   a) 40%   **b)** 36%   **c)** 15%

   d) 21%   **e)** 6%

5 At a local derby soccer match $\frac{17}{20}$ of the crowd were home supporters. What percentage was this?

6 In a survey about the colour of cars, 22% said they preferred red cars and $\frac{3}{20}$ said they preferred silver cars. Which of the two colours was more popular?

7 Put these numbers in order, smallest first.

   $\frac{3}{5}$ 30%   0·7 $\frac{3}{4}$ $\frac{2}{3}$

Exercise 16.1A cont'd

8 Write these percentages as decimals.

   a) 20%   **b)** 85%

   c) 41%   **d)** 6%

   e) 94·5%.

9 Write three fractions that are the same as 40%.

10 At Bradway school, $\frac{3}{5}$ of the teachers are female. What percentage is this?

EXERCISE 16.1B

1 Change the following fractions to decimals:

   a) $\frac{9}{10}$ **b)** $\frac{3}{8}$ **c)** $\frac{4}{5}$

   d) $\frac{17}{20}$ **e)** $\frac{3}{50}$

2 Change the decimals you found in question 1 to percentages.

3 Change the following fractions into percentages. Give the answers to 1 decimal place.

   a) $\frac{2}{3}$ **b)** $\frac{1}{6}$ **c)** $\frac{7}{8}$

   d) $\frac{17}{40}$ **e)** $\frac{9}{16}$

4 Change the following percentages to decimals:

   a) 47%   **b)** 31%   **c)** 11%

   d) 41%   **e)** 2%

5 Josh said that $\frac{4}{7}$ of the people in his class were boys. What percentage was this?

   Give the answer to 1 decimal place.

6 David walks $\frac{7}{8}$ of a kilometre to school and Paula walks 0·87 km to school. Who walks further?

7 Imran gives his mother $\frac{2}{9}$ of his weekly wage for food etc. What percentage is this?

   Give the answer to 1 decimal place.

8 Write these percentages as decimals.

   a) 30%   **b)** 15%   **c)** 23%

   d) 7%   **e)** 42·3%.

9 Write three fractions that are the same as 60%.

10 Put these numbers in order, smallest first.

   $\frac{3}{8}$ 35%   0·45 $\frac{2}{5}$ $\frac{5}{12}$

Finding fractions and percentages

To find a fraction of an amount, multiply the amount by the numerator and divide by the denominator.

You can be asked to do this either with or without a calculator.

> ### Exam tip
>
> For the non-calculator paper it is worth learning some basic equivalents.
>
> $\frac{1}{2}$ = 0·5 = 50% $\frac{1}{3}$ = 0·333... = 33·3...% $\frac{1}{5}$ = 0·2 = 20%
>
> $\frac{1}{4}$ = 0·25 = 25%   (33% to the nearest 1%) $\frac{2}{5}$ = 0·4 = 40%
>
> $\frac{3}{4}$ = 0·75 = 75% $\frac{2}{3}$ = 0·666... = 66·6...% $\frac{3}{5}$ = 0·6 = 60%
>
>   (67% to the nearest 1%) $\frac{4}{5}$ = 0·8 = 80%
>
> Notice the rounding of $\frac{1}{3}$ and $\frac{2}{3}$. It is a common error to assume $\frac{1}{3}$ = 0·3 = 30% and $\frac{2}{3}$ = 0·66 = 66% or even $\frac{2}{3}$ = 0·6 = 60%.

EXAMPLE 2

At a meeting there were 140 people. $\frac{2}{5}$ of them were men.

How many men were there?

a) Non calculator,

$\frac{2}{5} \times 140 = 140 \times 2 \div 5 = 280 \div 5 = 56.$

b) Calculator, work out $140 \times 2 \div 5$ which gives 56.

EXAMPLE 3

Without a calculator work out $\frac{3}{4}$ of 17.

Work out $17 \times 3 \div 4 = 51 \div 4 = 12\frac{3}{4}$

$51 \div 4 = 12$ and 3 left over $= 12\frac{3}{4}$

To find a percentage of an amount, it is easier to change to a decimal and then just multiply, either using a calculator or by normal multiplication.

EXAMPLE 4

Without using a calculator, work out 20% of £65.

20% of £65 = 0·2 × 65 = £13·00

EXAMPLE 5

David spends 23% of his income on his mortgage. He earns £31 000 a year. How much does he spend on his mortgage?

23% of 31 000 = 0·23 × 31 000 = £7130.

EXERCISE 16.2A

1 Work out $\frac{1}{5}$ of 20 m.
2 Work out 40% of £80.
3 Work out $\frac{5}{8}$ of 56 cm.
4 Work out 15% of £60.
5 Work out $\frac{3}{5}$ of 12.
6 Gayti spent 20% of her pocket money on magazines. She received £8 pocket money. How much did she spend on magazines?
7 At the Dronfield Brass Band concert $\frac{1}{3}$ of the audience were under 21. There were 180 people at the concert. How many were under 21?
8 Bryan bought a bike for £65. When he sold it he made a loss of 5%. What loss did he make?

9

The makers of 'Tomkat' cat food reduced the price of the standard tin to $\frac{5}{6}$ of the old price. The old price was £1·80. What was the new price?

10 Rod gives away 8% of his income. His monthly income is £1200. How much does he give away?

EXERCISE 16.2B

You may use a calculator in this exercise. Where answers are not exact, give them to the nearest penny or to 2 decimal places.

1 Work out $\frac{3}{5}$ of £720.
2 Work out 37% of 950 m.
3 Work out $\frac{5}{9}$ of 564.
4 Work out 95% of £172.
5 Work out $\frac{3}{20}$ of 15.
6 Selina paid 23% of her income in tax. How much tax did she pay in the month when she earned £1284?

7 Beardalls' jewellers offered $\frac{3}{8}$ off the price of all rings. How much did they offer off a ring priced at £420.

8 In the mix and match paint colours, $\frac{4}{5}$ of the paint used is white. To make 6 litres of paint how many millilitres of white paint is used?

Exercise 16.2B cont'd

9 At Park school 56% of the students in year 11 gained five or more good G.C.S.E.s. There are 125 pupils in the year. How many gained five or more good G.C.S.E.s?

10

Birkdale coach firm claims that over 85% of their seats in their coaches are fitted with safety belts. If this is true of a 53-seater coach, what is the least number of seats that must have seat belts fitted?

Key ideas

- To change a fraction into a decimal, divide the numerator by the denominator.
- To change a decimal to a percentage multiply by 100.
- To find a fraction of a quantity, multiply by the numerator and divide by the denominator.
- To find a percentage of a quantity, change the percentage to a decimal and just multiply.

Revision exercise

1 Sketch the plan and elevations of the following solid shapes.

The arrows indicate the direction of the plan, P, the front elevation, F, and the side elevation, S.

a)

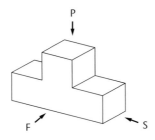

b)

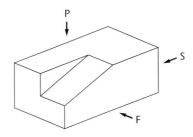

2 Draw accurately the plan and elevations of the following solid shape.

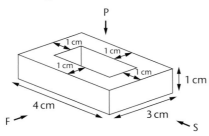

3 Two machines pack paper clips into boxes. Each box should hold 200 paper clips. A sample of 100 boxes was checked and the results are given in the table.

Machine A							
Number of paper clips	197	198	199	200	201	202	203
Frequency	1	8	28	27	22	10	4

Machine B								
Number of paper clips	197	198	199	200	201	202	203	204
Frequency	2	4	8	23	29	25	6	3

For each machine calculate the mean, median, mode and range.

Comment on your findings.

4 Draw axes from 0 to 8 for both x and y.

a) Plot the points $(1, 5)$, $(2, 5)$ and $(3, 8)$. Join to make a triangle and label it A.

b) Plot the points $(4, 1)$, $(5, 1)$ and $(6, 4)$. Join to make a triangle and label it B.

c) Describe fully the transformation that maps triangle A onto triangle B.

5

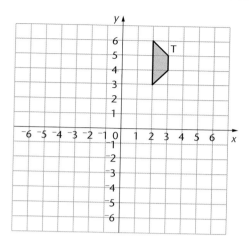

a) Translate the trapezium T, 3 to the right and 4 down. Label it A.

b) Translate the trapezium A, 2 to the left and 3 down. Label it B.

c) Translate the trapezium B, 8 to the left and 5 up. Label it C.

d) What movement is needed to translate C onto T?

6 Change these fractions to decimals, giving your answer to three decimal places where appropriate.

a) $\frac{7}{16}$ **b)** $\frac{3}{7}$ **c)** $\frac{9}{40}$ **d)** $\frac{11}{15}$.

7 Change the decimals you found in question 7 into percentages. Give your answers to one decimal place.

8 Do not use a calculator in this question.

a) Work out $\frac{1}{5}$ of £40

b) Work out 40% of £30.

9 John was earning £160 a week. He was given a rise of 12%. How much was his rise?

10 In 1966, 53% of households in a village had a black and white television. In the village there were 786 households. How many of them had a black and white television? Give the answer to the nearest whole number.

11 In Ponderosa Avenue $\frac{17}{20}$ of the houses have a car. There are 180 houses in the avenue. How many have a car?

17 Rotations

You should already know

- the meanings of line and rotational symmetry
- how to plot and read coordinates of a point

To rotate a shape, turn it around a point.

The shape is still the same way round, and it is congruent to the original shape.

For example this shape could turn to any of these positions

To rotate a shape you need to know three things

- the angle of rotation
- the direction of rotation
- the centre of rotation.

Remember that rotations are anticlockwise, unless you are told otherwise.

Tracing paper is useful both to draw rotations and to find the angle of turn when the rotation is given.

The easiest rotations are

- quarter-turns (90° anticlockwise)
- half-turns (180°)
- three-quarter-turns (90° clockwise)

about the centre of the shape or the origin.

In this diagram the shape P has been rotated through a quarter-turn (90° anticlockwise) about point O.

You can do this by counting squares, or by tracing the shape and turning it, or by using a pair of compasses with the point at O and drawing quarter circles from each point.

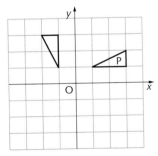

Exam tip

When you have drawn the rotation, turn the page through the correct angle to check it looks like the original.

EXAMPLE 1

Rotate the shape through a three-quarter-turn about O.

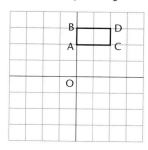 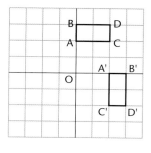

You can draw the rotation by counting squares.

A is 2 squares above O, so its image A' is 2 squares to the right.

B is 3 squares above O, so its image B' is 3 squares to the right.

C is 2 squares above and 2 to the right of O, so its image C' is 2 to the right and 2 below O.

D is 3 squares above and 2 to the right of O, so its image D' is 3 to the right and 2 below O.

You can think of a three-quarter-turn (anticlockwise) as a quarter-turn clockwise.

Again, you could do it by tracing.

EXERCISE 17.1A

Draw the diagrams on squared paper in this exercise.

1 Make three copies of this diagram and answer each part on a separate diagram.

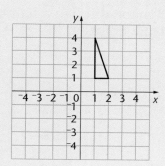

   a) Rotate this shape through a half-turn about the origin.

   b) Rotate this shape through 90° clockwise about the origin.

   c) Rotate this shape through 90° anticlockwise about the origin.

Exercise 17.1A cont'd

2 Make three copies of this diagram and answer each part on a separate diagram.

    a) Rotate this shape through 90° (anticlockwise) about the origin.

    b) Rotate this shape through 90° clockwise about the origin.

    c) Rotate this shape through 180° about the origin.

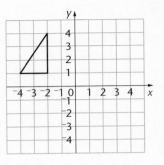

3 Make three copies of this diagram and answer each part on a separate diagram.

    a) Rotate this shape through 90° clockwise about the origin.

    b) Rotate this shape through 180° about the origin.

    c) Rotate this shape through 90° anticlockwise about the origin.

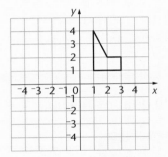

4 Rotate this shape through a half-turn about the point (4, 3).

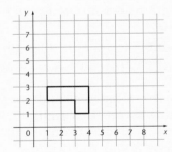

5 Rotate this shape through 90° (anticlockwise) about the point (5, 2).

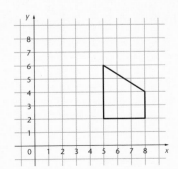

EXERCISE 17.1B

Draw the diagrams on squared paper in this exercise.

1 Make three copies of this diagram and answer each part on a separate diagram.

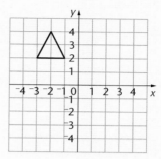

  a) Rotate this shape through a half-turn about the origin.

  b) Rotate this shape through 90° clockwise about the origin.

  c) Rotate this shape through 90° anticlockwise about the origin.

2 Make three copies of this diagram and answer each part on a separate diagram.

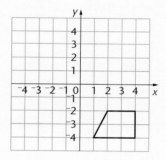

  a) Rotate this shape through 90° (anticlockwise) about the origin.

  b) Rotate this shape through 90° clockwise about the origin.

  c) Rotate this shape through 180° about the origin.

3 Make three copies of this diagram and answer each part on a separate diagram.

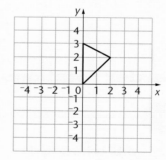

  a) Rotate this shape through 90° clockwise about the origin.

  b) Rotate this shape through 180° about the origin.

  c) Rotate this shape through 90° anticlockwise about the origin.

Exercise 17.1B cont'd

4 Rotate this shape through a half-
turn about the point (5, 3).

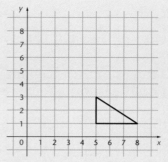

5 Rotate this shape through
90° (anticlockwise) about the
point (4, 4).

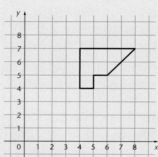

Exam tip

Tracing paper is always stated as optional
extra material in examinations. When doing
transformation questions, always ask for it.

Drawing harder rotations

EXAMPLE 2

Rotate triangle ABC through 90° clockwise about C.

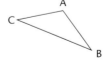

Measure an angle of 90° clockwise from the line AC.

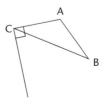

Example 2 cont'd

Trace the shape ABC. Put a pencil or pin at C to hold the tracing to the diagram at that point. Rotate the tracing paper until AC coincides with the new line you have drawn. Use another pin or the point of your compasses to prick through the other corners (A and B).

Join up the new points to form the image.

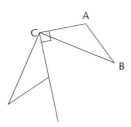

If the centre of rotation is not on the object then the method is slightly more difficult.

EXAMPLE 3

Rotate the triangle ABC through 90° clockwise about the point O.

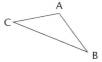

Join O to a point on the object (C).

Measure an angle of 90° clockwise from OC and draw a line.

Trace the triangle ABC and the line OC. Rotate the tracing about O until the line OC coincides with the new line you have drawn. Use a pin or the point of your compasses to prick through the other corners (A, B and C). Join up the pin holes to form the image.

For other angles of rotation (e.g. 120° clockwise), the first angle is measured as 120° not 90° but otherwise the method is identical.

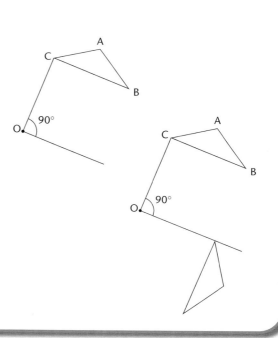

If the centre of rotation is easy to work with, for example the origin, then you may be able to draw the rotation by counting squares but it is always best to check using tracing paper.

Recognising rotations

It is usually easy to recognise when a transformation is a rotation, as it should be possible to place a tracing of the object over the image without turning the tracing paper over.

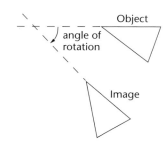

To find the angle of rotation, find a pair of sides that correspond in the object and the image. Measure the angle between them. You may need to extend both of these sides to do this.

If the centre of rotation is not on the object, its position may not be obvious. The easiest method to use is trial and error, either by counting squares or using tracing paper. In a later chapter, you will learn a method which will find the centre directly, without trial (or error!).

Exam tip

Always remember to state whether the rotation is clockwise or anticlockwise.

EXAMPLE 4

Describe fully the transformation that maps flag A onto flag B.

It is clear that the transformation is a rotation and that the angle is 90° clockwise. You may need to make a few trials, using tracing paper and a compass point centred on different points, to find that the centre of rotations is (7, 4).

If you did not spot it, try it now with tracing paper.

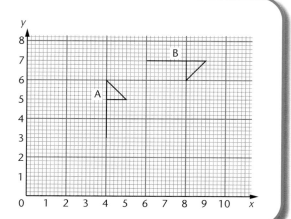

1 **a)** Rotate the shape through 90°, clockwise about (2, 2). Label it A.

   b) On the same grid rotate the first shape 90° anticlockwise about (1, 1). Label it B.

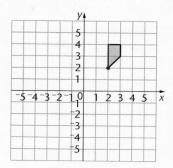

2 **a)** Rotate the shape through 180°, about (3, 2). Label it A.

   b) On the same grid rotate the first shape 90° anticlockwise about (1, 1). Label it B.

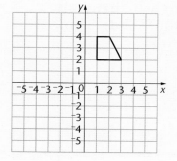

3 **a)** Rotate the shape through 90°, anticlockwise about (1, 3). Label it A.

   b) On the same grid rotate the first shape 180° about (1, ⁻2). Label it B.

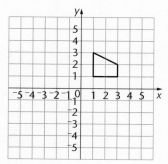

Exercise 17.2A cont'd

4 a) Rotate the shape through 90°, clockwise about (0, 0). Label it A.

b) On the same grid rotate the first shape 90° anticlockwise about (⁻2, 1). Label it B.

c) Describe fully the transformation that maps A onto B.

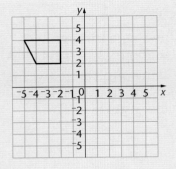

5 For the diagram describe fully the transformation that maps

a) A onto B   **b)** A onto C
c) A onto D   **d)** B onto E
e) D onto F.

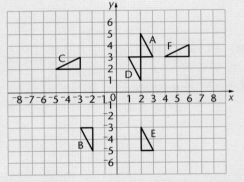

Exam tip

When describing transformations, always state the type of transformation first and then give all the necessary extra information. For translations this is the movement horizontally and vertically, for rotations it is angle, direction and centre. You will get no marks unless you actually name the transformation.

EXERCISE 17.2B

1 **a)** Rotate the shape through 90°, clockwise about (2, 2). Label it A.

   b) On the same grid rotate the first shape 90° anticlockwise about (1, 1). Label it B.

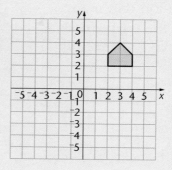

2 Rotate the triangle by 90° anticlockwise about the point (2, 0).

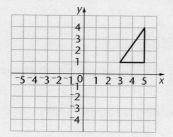

3 **a)** Rotate the shape through 90°, clockwise about 0. Label it A.

   b) Rotate A through 90° clockwise about 0. Label it B.

   c) Rotate B to complete a symmetrical pattern. Label it C.

   d) Describe fully the transformation that maps A onto C.

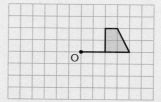

4 On squared paper mark axes ⁻5 to 5 for both x and y.

   a) Plot the points (⁻2, 1), (⁻2, 4), (⁻4, 3). Join up to form a triangle. Label it T.

   b) Rotate T through 90° anticlockwise about (⁻1, 0). Label it A.

   c) Rotate T through 90° clockwise about (0, 1). Label it B

   d) Describe fully the transformation that maps A onto B.

Exercise 17.2B cont'd

5 For the diagram describe fully the transformation that maps

a) A onto B   **b)** B onto C
c) B onto D   **d)** E onto F
e) A onto G.

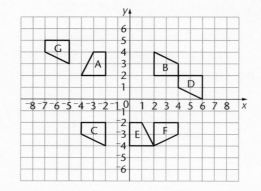

Key ideas

- In a rotation all the points move through the same angle in the same direction.
- In a rotation all the points remain at the same distance from the centre of rotation.
- When describing a rotation, you need the centre, the angle and the direction (anticlockwise is positive).

18 Expressing one quantity as a percentage of another

You should already know

- how to change a fraction to a percentage.

To express one quantity as a percentage of another, start by writing the first quantity as a fraction of the second.

Then use the methods already described to change the fraction to a percentage.

ACTIVITY 1

Write these fractions as percentages.

a) $\frac{1}{4}$ **b)** $\frac{1}{5}$ **c)** $\frac{3}{4}$

d) $\frac{3}{5}$ **e)** $\frac{7}{10}$

Exam tip

To do this both quantities must be in the same units.

EXAMPLE 1

Express 4 as a percentage of 5.

First write 4 as a fraction of 5 $\frac{4}{5}$

Then change $\frac{4}{5}$ to a percentage $\frac{4}{5} = 0.8 = 80\%$.

EXAMPLE 2

Express £5 as a percentage of £30.

$\frac{5}{30} = 0.167 = 16.7\%$ to one decimal place.

145

EXAMPLE 3

Express 70 cm as a percentage of 2·3 m.

First change 2·3 m to centimetres

2·3 m = 230 cm

$\frac{70}{230}$ = 0·304 = 30·4% to one decimal place

or

Change 70 cm to metres

70 cm = 0·7 m

$\frac{0·7}{2·3}$ = 0·304 = 30·4% to one decimal place.

EXERCISE 18.1A

1 Find

   a) 12 as a percentage of 100

   b) 4 as a percentage of 50

   c) 80 as a percentage of 200

   d) £5 as a percentage of £20

   e) 4 m as a percentage of 10 m

   f) 30 pence as percentage of £2

2 In form 7B 14 out of the 25 students are boys. What percentage is this?

3 A shopkeeper puts up a jumper which cost £20 by £1·50. What percentage increase is this?

4 Peter changed his gas supplier. It used to cost him £25 a month, but this has been reduced by £3. What percentage reduction is this?

5 Cindy used to deliver 40 newspapers. After a sales drive the number she delivered increased by 6. What percentage increase is this?

6 Annie scored 14 out of 20 in a test. What percentage was this.

7 Thelma spend £4 out of her £20 spending money on a visit to the theatre. What percentage was this?

8 At the Keep-fit club 8 out of the 25 members were men. What percentage was this?

9 In the Mayflower Hotel there are 15 double rooms and 5 single rooms. What percentage of the rooms are double?

10 Telford United won 19, lost 15 and drew 6 of the games in a season. What percentage of the games did they draw?

EXERCISE 18.1B

1 Work out, give the answers either exact or to one decimal place.

   a) 25 as a percentage of 200
   b) 40 as a percentage of 150
   c) 76 pence as a percentage of £1·60
   d) £1·70 as a percentage of £2
   e) 19 as a percentage of 24
   f) 213 as a percentage of 321

2 Penny scored 17 out of 40 in a test. What percentage was this?

3 Salma was given a discount of £2·50 on a meal that would have cost £17. What percentage discount had she been given?

4 Softa Suds used to sell a medium packet of soap powder that held 450g. They increased the contents by 50g. What percentage increase was this?

5 David won £7500 in a lottery. He gave £2300 of it to his son. What percentage was this?

6 At the Beauchief Theatre 184 out of the 240 people in the audience were women. What percentage of the audience were women? Give the answer to the nearest whole number.

7 During the 30 days of November in 2000, it rained on 28 of the days. On what percentage of the days did it rain? Give the answer to one decimal place.

8 In a radar speed control, 45 out of 167 motorists checked were found to be exceeding the speed limit. What percentage were exceeding the speed limit? Give the answer to one decimal place.

9 At the local derby between Sheffield United and Sheffield Wednesday there were 6198 United supporters and 4232 Wednesday supporters. What percentage of the crowd supported United? Give the answer to one decimal place.

10 Holly gets £5 a week from her grandad, £7 from her dad, £3 from her mum and earns £11 from a paper round. What percentage of her weekly income does she earn? Give the answer to one decimal place.

Key idea

● To find one quantity as a percentage of another, write the first quantity as a fraction of the second. Then change to a percentage. (Both quantities must be in the same units.)

19 Sequences

You should already know

- how to add, subtract, multiply and divide numbers
- how to substitute numbers for letters.

ACTIVITY 1

Look at these patterns.

In each case draw the next pattern. Then complete the table.

a)

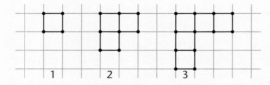

Pattern	1	2	3	4	5
Number of squares	1	2			
Number of dots	4	6			

b)

Pattern	1	2	3	4	5
Number of squares	1	3	5		
Number of dots	4	8			

Activity 1 cont'd

c)

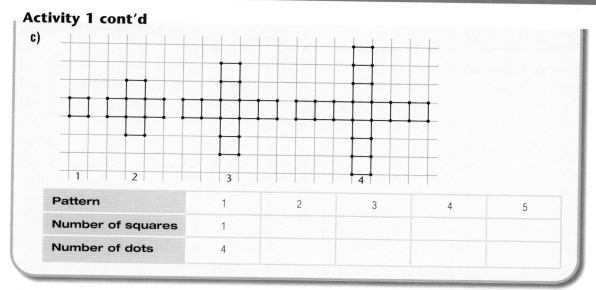

Pattern	1	2	3	4	5
Number of squares	1				
Number of dots	4				

Sequences

A group of numbers, written in order, is called a sequence if there is a rule to find the next number. The numbers in a sequence are called terms.

The rule is 'add 1' and the next term is 9.

The rule is 'multiply by 2' and the next term is 16.

When you are given only a few terms, you can sometimes spot more than one rule.

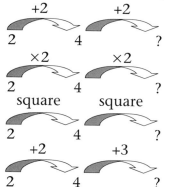

The rule is 'add 2,' and the next term is 6.

The rule is 'multiply by 2', the next term is 8.

The rule is 'square', the next term is 16.

The rule is 'add one more', the next term is 7. Can you find any more?

EXERCISE 19.1A

For each sequence, find a rule and write down the next number.

1	1	2	3	4
2	2	4	6	8
3	41	43	45	47
4	10	20	30	40
5	2	4	8	16
6	5	5	5	5
7	3	9	27	81
8	27	25	23	21
9	10	7	4	1
10	72	36	18	9

EXERCISE 19.1B

For each sequence, find a rule and write down the next number.

1	$\frac{1}{2}$	$\frac{1}{3}$	$\frac{1}{4}$	$\frac{1}{5}$
2	1	3	6	10
3	3	9	81	6561
4	64	32	16	8
5	1·2	1·5	1·8	2·1
6	243	81	27	9
7	3	7	13	21
8	1	2	6	15
9	2	8	18	32
10	1	1	2	3

The nth term

It is often possible to find a formula to give the terms in a sequence.

You usually use n to stand for the number of a term.

EXAMPLE 1

If the formula is nth term $= 2n + 1$, then:

the first term $= 2 \times 1 + 1 = 3$

the second term $= 2 \times 2 + 1 = 5$

the third term $= 2 \times 3 + 1 = 7$.

EXERCISE 19.2A

Write the first four terms of these sequences.

1 Starts with 1 and 2 is added each time.
2 Starts with 5 and it is doubled each time.
3 Starts with 16 and 3 is subtracted each time.
4 Starts with 40 and it is halved each time.

For the rest of the questions the nth term is given. Write the first four terms of each of these sequences.

5 $n + 2$
6 $3n$
7 $2n - 1$
8 $2n + 3$
9 $6n - 5$
10 $n - 2$

EXERCISE 19.2B

Write the first four terms of these sequences.

1 Starts with 2 and 3 is added each time.

2 Starts with 2 and it is multiplied by 3 each time.

3 Starts with 20 and 4 is subtracted each time.

4 Starts with 16 and it is halved each time.

For the rest of the questions the nth term is given. Write the first four terms of each of these sequences.

5 $n + 4$

6 $2n$

7 $3n - 1$

8 $4n + 3$

9 $5n - 3$

10 $2n - 5$

Key ideas

● Describe the rule in terms of add, subtract, multiply by or divide by something.
● To find the next term, work out the rule and use it.

E1 Revision exercise

1 a) On a grid labelled ⁻4 to 4 on both axes, plot the points (1, 1), (1, 3) and (2, 3) and join to make a triangle T.

b) Rotate triangle T through a half-turn about the origin.

2 a) On another grid draw the triangle in question 1 again.

b) Rotate triangle T through 90° anticlockwise about the origin.

3 a) On another grid draw the triangle in question 1 again.

b) Rotate triangle T through 180° about the point (1, 1).

4 a) On another grid draw the triangle in question 1 again.

b) Rotate triangle T through 90° anticlockwise about the point (⁻1, 1).

5

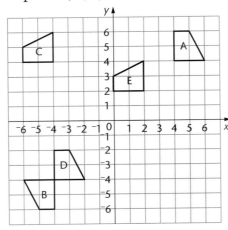

Describe fully the transformation that maps

a) A onto B   **b)** A onto C
c) B onto D   **d)** A onto E

6 Write

a) 14 as a percentage of 35
b) £48 as a percentage of £400
c) 20 cm as a percentage of 4 m

7 Promo washing powder used to cost £3·60 for a large packet. The price was increased by 27 pence. What percentage increase was this?

8 Find £2·50 as a percentage of £31. Give your answer to the nearest 1%.

9 Find 40 cm as a percentage of 3 m. Give your answer to one decimal place.

10 For each of these sequences, write down the rule and the next term.

a) 3	5	7	9	
b) 5	10	15	20	
c) 2	4	8	16	
d) 5	3	1	⁻1	
e) 32	16	8	4	

Stage 6

1 Using a calculator effectively

You should already know

- use a calculator to do simple calculations
- find the mean of a set of numbers.

Calculators vary in how they do calculations

For example if you use a calculator to work out $2 + 3 \times 4$ some will give the answer 20 because they work from 'left to right' that is $2 + 3 = 5$ then multiply the 5 by 4, 20 is the wrong answer; other calculators will give the answer 14 because these calculators follow the rules of arithmetic and do the multiplication first, that is $3 \times 4 = 12$ and then add 2. 14 is the correct answer.

Work through Activity 1 below.

ACTIVITY 1

Use a calculator to work these out
a) $13 \cdot 75 + (2 \cdot 25 \times 3)$
b) $14 \cdot 6 + 3 \times 9 \cdot 4$
a) $13 \cdot 75 + (2 \cdot 25 \times 3) = 13 \cdot 75 + 6 \cdot 75 = 20 \cdot 50$
b) $14 \cdot 6 + 3 \times 9 \cdot 4 = 14 \cdot 6 + 28 \cdot 2 = 42 \cdot 8$

ACTIVITY 2

Using only + and × and the given numbers try to make a sum to give the totals required.
a) use 3, 5 and 6 to get (i) 33 and (ii) 23
b) use 17, 11 and 5 to get (i) 72 and (ii) 96.

EXERCISE 1.1A

Use a calculator to work these out.
1 $23 \cdot 4 + 4 \times 5$
2 $19 \cdot 8 - 3 \cdot 2 \times 4$
3 $13 \cdot 75 - 2 \cdot 25 \times 3$
4 $14 \cdot 6 + 4 \times 1 \cdot 9$
5 $9 \cdot 8 + 9 \cdot 8 \times 9 \cdot 8$

EXERCISE 1.1B

Use a calculator to work these out.

1 $19 - 4 \times 2\cdot9$

2 $25\cdot5 - 5 \times 4\cdot9$

3 $37\cdot8 + 12 \times 1\cdot9$

4 $100 + 2\cdot2 \times 100$

5 $1000 - 25 \times 25$

More complicated calculations are made straightforward if you use brackets.

EXAMPLE 1

Use a calculator to work these out

a) $4\cdot8 \times 3\cdot7 - 2\cdot9 \times 2\cdot3$

b) $^-4\cdot6 \times 7\cdot2 + 3\cdot1 \times 4\cdot3$

c) $\dfrac{9\cdot7 + 4\cdot6}{3 \times 2}$

a) $4\cdot8 \times 3\cdot7 - 2\cdot9 \times 2\cdot3$

$= (4\cdot8 \times 3\cdot7) - (2\cdot9 \times 2\cdot3)$ put brackets around the multiplication

$= 17\cdot76 - 6\cdot67$

$= 11\cdot09$

Key in:

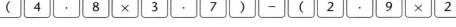

(4 · 8 × 3 · 7) − (2 · 9 × 2

· 3) =

b) $^-4\cdot6 \times 7\cdot2 + 3\cdot1 \times 4\cdot3$

$= (^-4\cdot6 \times 7\cdot2) + (3\cdot1 \times 4\cdot3)$ put brackets around the multiplication

$= ^-33\cdot12 + 13\cdot33$

$= ^-19\cdot79$

Key in:

(− 4 · 6 × 7 · 2) + (3 · 1 ×

4 · 3) =

c) $\dfrac{9\cdot7 + 4\cdot6}{3 \times 2}$

$= \dfrac{(9\cdot7 + 4\cdot6)}{(3 \times 2)}$ put brackets around the multiplication and the addition

$= \dfrac{14\cdot3}{6}$

$= 2\cdot383$

EXERCISE 1.2A

Use a calculator to work these out.

1 ⁻4·73 + 2·96 − 1·71 + 3·62

2 (⁻4·6 × 7·2) + (3·1 × ⁻4·3)

3 $\dfrac{^{-}4\cdot7 + 2\cdot6}{^{-}5\cdot7}$

4 $\dfrac{7\cdot92 \times 1\cdot71}{^{-}4\cdot2 + 3\cdot6}$

EXERCISE 1.2B

Use a calculator to work these out.

1 ⁻14·7 + 6·92 − 1·41 − 2·83

2 (⁻1·2 × ⁻2·4) − (9·2 × ⁻3·6)

3 $\dfrac{^{-}4\cdot72}{^{-}1\cdot4} \times \dfrac{8\cdot61}{^{-}7\cdot21}$

4 $\dfrac{3\cdot14 - 8\cdot16}{^{-}8\cdot25 \times 3\cdot18}$

EXAMPLE 2

Use a calculator to work these out

a) $\dfrac{14\cdot73 + 2\cdot96}{15\cdot25 - 7\cdot14}$ **b)** $\sqrt{17\cdot8^2 + 4\cdot3^2}$.

a) $\dfrac{14\cdot73 + 2\cdot96}{15\cdot25 - 7\cdot14}$ = 2·1812 = 2·18 (to 2 d.p.)

There are various ways to do this. One is to key in:

| 1 | 4 | · | 7 | 3 | + | 2 | · | 9 | 6 | = |

| ÷ | (| 1 | 5 |

| · | 2 | 5 | − | 7 | · | 1 | 4 |) | = |

b) $\sqrt{17\cdot8^2 + 4\cdot3^2}$ = 18·312 = 18·3 (to 3 s.f.)

Key in: | (| 1 | 7 | · | 8 | x^2 | + | 4 | · | 3 | x^2 |) | √ | = |

Some of the more common operations you will be asked to use are powers and roots (using the $\boxed{y^x}$ button and the $\boxed{^x\sqrt{}}$ or $\boxed{y^{1/x}}$ button).

Exam tip

Where it is possible to split the question up, it is useful to do so to check, but make sure you do not round the answer too early. Unless the question states otherwise it is best to give your final answer to 3 s.f.

EXAMPLE 3

Use a calculator to evaluate these

a) $4 \cdot 2^3$ **b)** $^5\sqrt{15}$.

a) $4 \cdot 2^3 = 74 \cdot 088 = 74 \cdot 1$ (to 3 s.f.)

This can be done by keying $4 \cdot 2 \times 4 \cdot 2 \times 4 \cdot 2$, but using the y^x button is quicker.

Key in: | 4 | · | 2 | y^x | 3 | = |

b) $^5\sqrt{15} = 1 \cdot 7187 = 1 \cdot 72$ (to 3 s.f.)

Key in: | 5 | 2nd F | y^x | 1 | 5 | = |

This may vary, depending on your calculator. $^5\sqrt{15}$ is the same as $15^{\frac{1}{5}}$.

EXERCISE 1.3A

Work these out.

1 **a)** $3 \cdot 8^4$ **b)** $^4\sqrt{31 \cdot 8}$

2 $43 \cdot 73^3 + 17 \cdot 1^2$

3 $\sqrt{7^2 + 14^2}$

4 $(9 \cdot 2 + 15 \cdot 3)^2$

5 $\dfrac{6 \cdot 2}{2 \cdot 6} + \dfrac{5 \cdot 4}{3 \cdot 9}$

6 $\dfrac{2 \cdot 6 + 4 \cdot 25}{7 \cdot 8 \times 3 \cdot 6^2}$

EXERCISE 1.3B

Work these out.

1 **a)** $7 \cdot 31^5$ **b)** $^5\sqrt{12 \cdot 2}$

2 $7^2 + 3 \cdot 6^2 - 2 \times 2 \cdot 7 \times 3 \cdot 6 \times 0 \cdot 146$

3 $\dfrac{9 \cdot 4}{4 \cdot 31} - \dfrac{2 \cdot 6}{5 \cdot 27}$

4 $\left(2 \cdot 4 \times 3 \cdot 1 - \dfrac{6 \cdot 8}{3 \cdot 4}\right)^2$

5 $0 \cdot 5 \,(^-5 + \sqrt{5^2 + 1200})$

6 $\dfrac{19 \cdot 4 \times 6 \cdot 3 - 2 \cdot 61}{8 \cdot 1 + 7 \cdot 94}$

You need to be able to use your calculator to answer questions involving measures. This is straightforward with measures such as length, weight etc. because they use a decimal system, provided, that is, you take care with the units.

With time however you need to remember that there are 60 seconds in a minute and 60 minutes in an hour – time is not a decimal system. Thus you may find it easier for example to add minutes and seconds separately.

EXAMPLE 4

The weight of 8 bags of potatoes are

5 kg 800 g, 5 kg 200 g, 6 kg 500 g, 4 kg 900 g, 4 kg 750 g, 6 kg 100 g, 5 kg 600 g, 5 kg.

Find the mean weight.

Hint: convert the weights to decimals as you enter them.

Answer: $\dfrac{5 \cdot 8 + 5 \cdot 2 + 6 \cdot 5 + 4 \cdot 9 + 4 \cdot 75 + 6 \cdot 1 + 5 \cdot 6 + 5}{8}$

$= \dfrac{43 \cdot 85}{8}$

$= 5 \cdot 48$ kg or 5 kg 480 g.

EXAMPLE 5

The times of four children in a relay race were
9 min 10 s, 9 min 43 s, 9 min 49 s, 9 min 53 s.
Find the total time the race took.

Answer   The total time = (9 + 9 + 9 + 9) minutes + (10 + 43 + 49 + 53) seconds

= 36 min + 155 s

= 36 min + 2 min + 35 s

= 38 min 35 s.

EXERCISE 1.4A

1 Find the total weight of eight sacks of gravel weighing:

25 kg 750 g, 26 kg 800 g, 30 kg 7 g, 29 kg 85 g, 29 kg 430 g, 28 kg 106 g, 26 kg 200 g, 25 kg 400 g.

2 The time that the first seven patients spent with a doctor one morning between 9:00 a.m. and 10:00 a.m. were:

3 min 20 s, 5 min 35 s, 1 min 3 s, 4 min exactly, 2 min 50 s, 4 min 40 s, 6 min 25 s.

   a) What was the total time?
   b) How much time did the doctor have left before 10:00 a.m.

3 Find the mean weight of four children whose weights are:

36 kg 600 g, 29 kg 840 g, 43 kg 130 g, 44 kg 436 g.

4 The total weight of seven girls is 374 kg and the total weight of five boys is 314 kg. Calculate the mean weight of all the children.

5 Six children estimated the length of a line. Their estimates were:

1 m 25 cm, 1 m 40 cm, 1 m 12 cm, 1 m 30 cm, 1 m, 1 m 10 cm

Find the mean of their guesses.

EXERCISE 1.4B

1 Find the total volume of water in a tank when six bottles of water were poured in if the volume of water in each bottle was:

2 litres 350 ml, 3 litres 100 ml, 2 litres 930 ml, 5 litres 255 ml, 3 litres 700 ml, 750 ml.

Exercise 1.4B cont'd

2 Six children cycled round a lake in a relay race.

Their times were 38 min 20 s, 41 min 10 s, 40 min 40 s, 39 min 39 s, 43 min 32 s, 42 min 18 s.

How long did the race last?

3 The sign in a lift states:

'Maximum weight 800 kg'

If five people weighing

78 kg, 74 kg 100 g, 70 kg 350 g, 69 kg 850 g, 72 kg 360 g

are in the lift how many kilograms short of the maximum weight is their total weight?

4 Tony weighs his three cats. The weights are 5·3 kg, 6·2 kg, 6·1 kg. Find the mean weight.

5 Jodi keeps a record of how long it takes her to drive from her home to work each morning for a week. Here are the times:

45 min, 36 min, 34 min, 42 min, 32 min.

Find the mean time for her journey.

Key ideas

- Use the brackets function on your calculator to make calculations more straightforward.
- Use the y^x and the second function keys to simplify calculations with powers and roots.

Brackets and factors

You should already know

- **how to calculate with negative numbers**
- **how to calculate squares of numbers**
- **how to substitute numbers for letter.**

Brackets

What is $2 \times 3 + 4$? Is it 14? Is it 10?

The rule is 'do the multiplication first', so the answer is 10.

If you want the answer to be 14, you need to add 3 and 4 first.

Use brackets to show this $2 \times (3 + 4)$.

Notice that this is equal to $2 \times 3 + 2 \times 4$.

ACTIVITY 1

1 You may have used brackets when finding the areas of shapes and/or when learning about multiplication.
For example

The area is $2 \times (3 + 4) = 2 \times 3 + 2 \times 4$
$= 6 + 8$
$= 14.$

In the same way write down the areas of the following shapes **a)** using brackets and then **b)** finding the area of the whole shape.

a)

b)
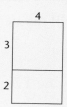

c)
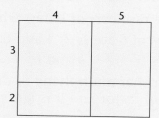

d)

2 Draw diagrams to show these expressions

a) $4(6 + 2)$ **b)** $a(c + d)$ **c)** $y(y + 1)$

It is the same in algebra.

$a(a + b)$ means 'add b and c then multiply by a' and this is the same as 'multiply a by b, multiply b by c, then add the results'.

So $a(b + c) = ab + ac$.

Exam tip

This is called expanding the brackets. Remember to multiply the number or letter outside by each term inside the brackets.

EXAMPLE 1

Expand the brackets.

a) $5(2x + 3) = 5 \times 2x + 5 \times 3$
$$= 10x + 15$$

b) $4(2x - 1) = 4 \times 2x + 4 \times (^-1)$
$$= 8x - 4$$

Exam tip

Remember about multiplying with negative numbers!
$$(^-3) \times (^-x) = 3x$$
$$(^-3) \times x = {}^-3x$$
and so on.

EXERCISE 2.1A

Expand these brackets.

1 $2(a + b)$ **6** $2(1 - x)$

2 $3(x + 2)$ **7** $5(p - q)$

3 $4(2x + 1)$ **8** $3(3x - 1)$

4 $a(a + 2)$ **9** $2(3x + 2)$

5 $y(y - 1)$ **10** $2(2x - 3)$

EXERCISE 2.1B

Expand these brackets.

1 $3(x + y)$ **6** $3c(c + 4)$

2 $2(p + 3)$ **7** $x(2 - x)$

3 $x(x + 3)$ **8** $^-y(2 + y)$

4 $4(3x - 2)$ **9** $^-z(z - 2)$

5 $a(a + 2)$ **10** $^-2x(5x - 3)$

Factorising algebraic expressions

Factors are numbers or letters which will divide into an expression.

The factors of 6 are 1, 2, 3 and 6.

The factors b^2 are 1, b and b^2.

Remember that multiplying or dividing by 1 leaves a number unchanged, so 1 is not a useful factor and it is ignored.

Factor trees can help. Here are 2 to remind you:

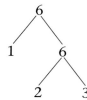

$1 \times 6 = 6$

$1 \times 2 \times 3 = 6$
ie $2 \times 3 = 6$

$1 \times 8 = 8$

$1 \times 2 \times 4 = 8$
ie $2 \times 4 = 8$

$1 \times 2 \times 2 \times 2 = 8$
ie $2 \times 2 \times 2 = 8$

To factorise an expression, look for common factors, for example, the common factors of $2a^2$ are $6a$ and 2, a and $2a$.

EXAMPLE 2

Factorise these fully.

a) $4p + 6$ **b)** $2a^2 - 3a$.

a) $4p + 6 = 2(2p + 3)$

The only common factor is 2 and $2 \times 2p = 4p$, $2 \times 3 = 6$.

b) $2a^2 - 3a = a(2a - 3)$

The only common factor is a and $a \times 2a = 2a^2$,
$a \times {}^-3 = {}^-3a$.

Exam tip

Make sure that you have found all the common factors. Check that the expression in the bracket will not factorise further.

EXERCISE 2.2A

1 Complete the following factorisations.

   a) $12a + 3 = 3(? + 1)$ **b)** $6b - 4 = ?(3b - 2)$ **c)** $4xy + 2yz = 2y(? + ?)$

Factorise these fully, where possible.

2 $5a^2 + 10b$ **5** $10x^2 - 100$ **7** $15z - 3$

3 $2x^2 + 3$ **6** $9a^2 - 27$ **8** $12q^2 - 18q$

4 $3p + 4p^2$

EXERCISE 2.2B

1 Complete the following factorisations.

a) $9a + 18 = 9(a + ?)$ **b)** $5y - 30 = 5(y - ?)$

c) $4x + 16 = ?(x + 4)$ **d)** $2ab + 6bc = 2b(? + ?)$

Factorise these fully.

2 $7m^2 - 49$ **5** $3p^2 - p$ **7** $x^2 + 7x$

3 $6 - 3a^2$ **6** $5 - 15x^2$ **8** $21y^2 - 7y$

4 $24 + 36a^2$

Key ideas

- Everything inside a bracket must be multiplied by the term outside the bracket.
- When finding common factors make sure you factorise fully

3 Angles

You should already know

- the language of shapes such as:
  triangle, vertex, parallel, quadrilateral
- the different types of triangle such as isosceles, equilateral and scalene.

Some basic angle facts

You may have already met and used many of the angle facts listed below.
Check that you know them all.

1 The sum of angles on a straight line is 180°.

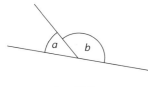

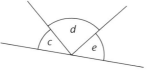

a + b = 180°   c + d + e = 180°

> **Exam tip**
>
> If you don't know these angle facts, learn them so that you can use them easily when needed.

2 The sum of angles round a point is 360°.

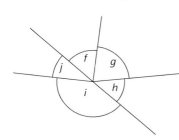

f + g + h + i + j = 360°

3 When two straight lines cross, the opposite angles are equal.

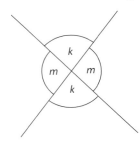

The angles *k* are equal. The angles *m* are equal. They are opposite angles at a point. Angles like these are also called vertically opposite angles. Also, $k + m = 180°$ as they are angles on a straight line.

4 The sum of the angles in a triangle is 180°.

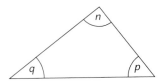

$n + p + q = 180°$

We will prove this fact later on in this chapter.

5 The sum of the angles in a quadrilateral is 360°.

 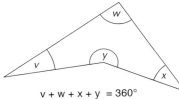 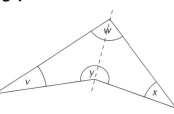

$r + s + t + u = 360°$ $\qquad$ $v + w + x + y = 360°$

The sum in the second quadrilateral is still 360°, even though it is re-entrant rather than convex in shape. This is easy to see if you split the quadrilateral into two triangles. Each triangle has an angle sum of 180°, so the sum of the angles of the quadrilateral is $2 \times 180° = 360°$.

EXAMPLE 1

Find the missing angles in these diagrams.

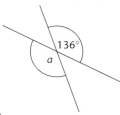

$a = 136°$
opposite angles at a point
are equal

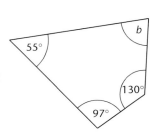

$55° + 97° + 130° = 282°$
So $b + 282° = 360°$
$\qquad b = 78°$

angle sum of a
quadrilateral is 360°

EXAMPLE 2

Find the value of *a* in this isosceles triangle.

Since the triangle is isosceles, the unmarked angles is also *a*.

$2a + 38° = 180°$

So $2a = 180° − 38°$

$2a = 142°$

$a = 71°$

angle sum of a triangle is 180°

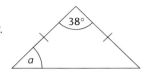

EXERCISE 3.1A

1 Two of the angles of a triangle are 30° and 72°. What is the size of the third angle of the triangle?

2 Find the sizes of these lettered angles.

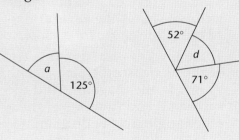

3 Two lines cross, making an angle of 35°. What are the sizes of the other three angles?

4 Two angles make a straight line. One of the angles is 87°. What is the other?

5 Three angles together make a straight line. Two of them are 52°. What is the other?

6 Find the sizes of the lettered angles, if AB and CD are straight lines.

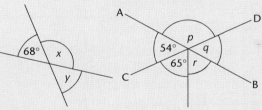

7 Look at these angles, then copy the table below and complete it.

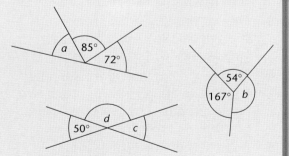

Chapter 3 *Angles*

Exercise 3.1A cont'd

Angle	Size	Reason
a	__°	angles on a straight line add to __°
b	__°	angles at a point add to __°
c	__°	… angles at a point are equal
d	__°	angles on a straight line …

8 Three angles of a quadrilateral are 100°. What is the size of the fourth angle?

9 A quadrilateral has two pairs of equal angles. One pair are each 55°. What size is each of the other two angles?

10 Draw a circle, then draw a quadrilateral with all its vertices on the edge of the circle. Measure the opposite angles of the quadrilateral. If your drawing is accurate, the opposite pairs should add to 180°.

EXERCISE 3.1B

1 Draw an isosceles triangle and its line of symmetry. Show that the line of symmetry splits the isosceles triangle into two congruent right-angled triangles.

2 One of the equal angles of an isosceles triangle is 35°. Find the other two angles.

3 One of the angles of an isosceles triangle is 126°. Find the other two angles.

4 One of the angles of an isosceles triangle is 50°. Find the other two angles. (Hint: There are two sets of possible answers.)

5 A circle is divided into six equal sectors. What is the angle at the centre of the circle, for each sector?

6 A circle is divided into nine equal sectors. What is the angle at the centre of the circle, for each sector?

7 Three lines cross at a point. Two of the angles formed are equal, as shown in the diagram. Find the value of a and the sizes of the other angles in the diagram.

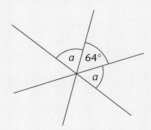

Exercise 3.1B cont'd

8 The two unmarked angles in this kite are equal. Find their size.

9 A kite has two angles of 125° and one angle of 75°. Find the size of its other angle.

10 The two unmarked angles in this arrowhead are equal. Find their size.

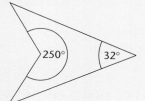

Angles with parallel lines

When a line crosses a pair of parallel lines, sets of equal angles are formed. The line cutting the parallels is called a **transversal**.

Learn to recognise each type of pair so that you are able to give reasons for angles being equal.

The diagrams above show pairs of equal alternate angles. These are on opposite sides of the transversal.

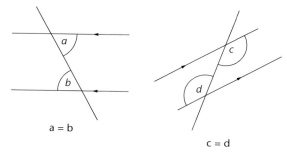

$a = b$

$c = d$

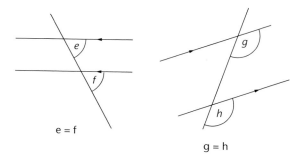

$e = f$

$g = h$

Exam tip

Thinking of a Z shape may help you to remember alternate angles are equal. Alternatively, turn the page round and look at the diagrams upside down and you will see the same shapes!

The diagrams above show pairs of equal corresponding angles. These are in the same position between the transversal and one of the parallel lines.

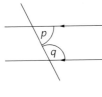

p + q = 180°

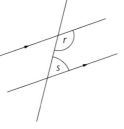

r + s = 180°

These pairs of angles are not equal. Instead, they add to 180°. They are called allied angles or co-interior angles.

EXAMPLE 3

Find the sizes of the lettered angles in these diagrams, giving your reasons.

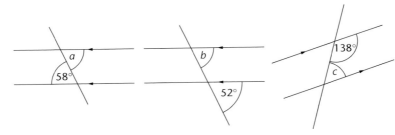

a = 58°   alternate angles are equal.
b = 52°   corresponding angles are equal.
c = 42°   allied angles add to 180°.

Notation for lines and angles

You often use single letters for angles and lines, as in this chapter so far. Sometimes, the ends of a line are labelled with letters. Then you can use these letters to name the angles.

The line AB is the line joining A and B.

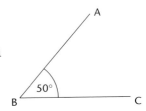

The angle at B is the angle made between the lines AB and BC. It may be written as angle ABC.

In this diagram, angle ABC = 50°.

It is essential to use three letters when there is more than one angle at a point.

In this diagram, there are four angles at O and they are two different sizes.

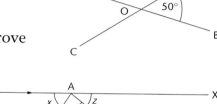

Angle AOC = angle BOD = 50°

Angle BOC = angle AOD = 130°

We can use the angle facts for parallel lines to prove that the sum of the angles in a triangle is 180°.

ABC is a triangle.

Line AX is parallel to line BC. Angle x + angle y + angle z = 180° because the sum of the angles on a straight line = 180°. Angle x = angle p because they are alternate angles. For the same reason angle z = angle q. Therefore angle p + angle q + angle y = 180°.

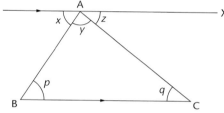

EXERCISE 3.2A

1 Find the value of x, giving your reason.

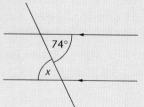

2 Find the value of z, giving your reason.

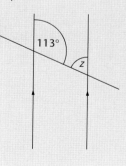

3 Draw accurately a pair of parallel lines and a line crossing them. Mark on your drawing a pair of acute corresponding angles. Measure them and check that they are equal.

4 Draw accurately a pair of parallel lines and a line crossing them. Mark on your drawing a pair of obtuse alternate angles. Measure them and check that they are equal.

5 Find the sizes of the lettered angles in this diagram.

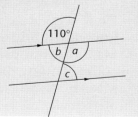

Exercise 3.2A cont'd

6 Find the sizes of the lettered angles in this diagram.

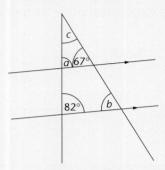

7 In a parallelogram ABCD, angle ABC = 75°. Draw a sketch and mark this angle and the sizes of the other three angles.

8 One of the angles of an isosceles trapezium is 64°. Draw a sketch and write the sizes of all the angles in suitable positions on your sketch.

9 Sketch an isosceles trapezium and write down all you can about its angles and symmetry.

10 Find the sizes of the lettered angles in this diagram.

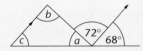

EXERCISE 3.2B

1 Find the value of *y*, giving your reason.

2 Draw accurately a pair of parallel lines and a line crossing them. Mark on your drawing a pair of acute alternate angles. Measure them and check that they are equal.

3 Draw accurately a pair of parallel lines and a line crossing them. Mark on your drawing a pair of corresponding obtuse angles. Measure them and check that they are equal.

4 Draw accurately a pair of parallel lines and a line crossing them. Mark on your drawing a pair of allied angles. Measure them and check that they add up to 180°.

5 Find the sizes of the lettered angles in the diagram.

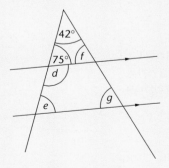

Exercise 3.2B cont'd

6 In a parallelogram, one angle is 126°. Make a sketch of the parallelogram and mark this angle. Calculate the other three angles and label them on your sketch.

7 In a isosceles trapezium ABCD, angle BCD is 127°. Make a sketch of the trapezium and mark this angle. Calculate the other three angles and label them on your sketch.

8 Sketch a parallelogram and write down all you can about its angles and its symmetry.

9 Find the sizes of the lettered angles in this diagram.

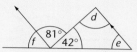

10 Use what you know about triangles to find the sizes of angles ABE and CDE in this diagram. Show why BE and CD are parallel.

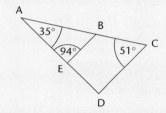

Exterior angle of a triangle

If one of the sides of a triangle is extended, to form an angle outside the triangle, this angle is called the exterior angle.

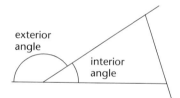

 exterior angle / interior angle

ACTIVITY 1

I) Draw a fairly large triangle with sides 10 cm, 12 cm, 14 cm and extend the length of the base.

Cut out or tear off two of the corners, as shown by the shaded angles and try to fit them alongside the base, in other words over the top of the exterior angles.

What do you notice?

2) Now repeat for a different sized triangle with a different exterior angle. Does the same thing happen?

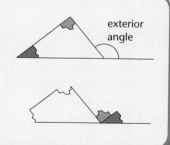

exterior angle

You have just demonstrated that the exterior angle of a triangle is equal to the sum of the interior opposite angles

However you can prove that the exterior angle equals the sum of the interior opposite angles as follows.

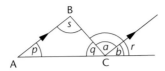

angle a + angle b + angle q = 180°   because angles on a straight line sum to 180°.
angle p = angle b because they are corresponding angles.
angle s = angle a because they are alternate angles.
angle p + angle s = angle a + angle b
angle a + angle b = angle r
angle p + angle s = angle r

EXERCISE 3.3A

1 Calculate the sizes of all the angles marked with letters.

a) x, 18°, 134°

b) 120°, b, 40°

c) a, 35°, 80°

d) 60°, 68°, c

2 Write down as many different ways as you can to find the value of e once you have found the value of d.

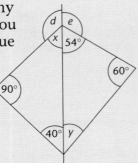

d, e, x, 54°, 60°, 90°, 40°, y

3 Calculate the sizes of all the angles marked with letters.

a) 60°, x, 28°

b) 135°, 75°, a

c) a, 40°, 80°

d)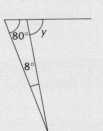
80°, y, 8°

Exercise 3.3A cont'd

4 Calculate the sizes of all the angles marked with letters.

a)

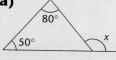

b)

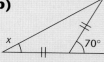

c)

d)

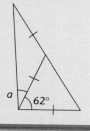

e)

EXERCISE 3.3B

1 Calculate the sizes of all the angles marked with letters.

a)

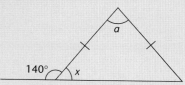

b)

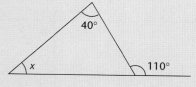

c)

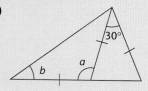

2 Calculate the sizes of all the angles marked with letters.

a)

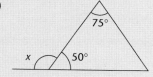

b)

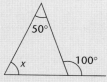

c)

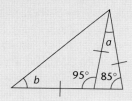

Exercise 3.3B cont'd

3 Calculate the sizes of all the angles marked with letters.

a)

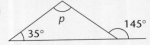

b)

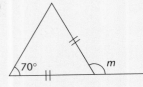

4 Calculate the sizes of all the angles marked with letters.

a)

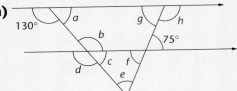

b)

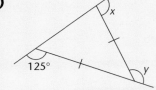

c)

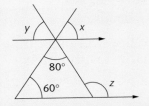

Angles in polygons

ACTIVITY 2

Draw a polygon with 5 sides. This is called a pentagon. Put a pencil along the base of the pentagon, as shown, then slide it to the right until its end reaches the end of the base. Now carefully rotate the pencil about its end until it is along the next side of the pentagon. Slide it up until its end reaches the top of that side. Continue in this way until you reach the base again, as shown. When you get back to the beginning, the pencil will have turned through 360°.

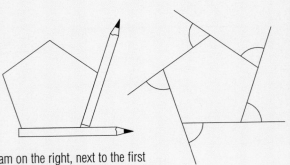

It will have turned through the angles shown in the diagram on the right, next to the first diagram. These angles are called exterior angles. You have shown that the sum of the exterior angles of the pentagon is 360°.

This method could have been used for any convex polygon.

> The sum of the exterior angles of any convex polygon is 360°.

At each vertex, the interior and exterior angles make a straight line.

> For any convex polygon, at any vertex: interior angle + exterior angle = 180°.

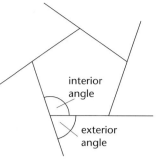

interior angle

exterior angle

EXAMPLE 4

The two unlabelled exterior angles of this pentagon are equal. Find their size.

First, find the sum of the angles that are given.

72° + 80° + 86° = 238°.

The sum of all the exterior angles is 360°.

The sum of the remaining two angles

= 360° − 238° = 122°.

So each angle is 122 ÷ 2 = 61°.

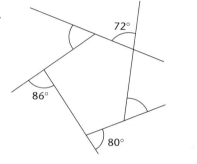

72°

86°

80°

In many problems about polygon angles, the easiest way to solve them is to use the fact that the sum of the exterior angle is 360°.

Sometimes, however, it is useful to find the sum of the interior angles.

At each vertex of a convex polygon: interior angle + exterior angle = 180°.

> Sum of (interior + exterior angles) for the polygon = 180° × number of angles
> = 180° × number of sides.

But the sum of the exterior angles is 360°.

So the sum of the interior angles of the polygon = 180° × number of sides − 360°.

Putting this algebraically, for an n-sided convex polygon:

the sum of the interior angles = $(180n - 360)°$.

EXAMPLE 5

Find the value of the missing angle in this pentagon.

For a pentagon:

the sum of the interior angles

$= (180 \times 5 - 360)° = 540°$.

The sum of the four angles given $= 437°$.

So the missing angle $= 540° - 437° = 103°$.

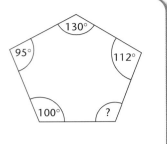

Finding angles in regular polygons

For regular polygons, each interior angle is the same, each exterior angle is the same. This is an easy way to work out the angles.

Another useful fact about regular polygons is that they may be divided, from their centre, into congruent isosceles triangles. This is often useful when you are asked to draw regular polygons accurately. For example, for a regular pentagon, the angle at the centre is $360° \div 5 = 72°$.

The base angles of the triangle together equal $180° - 72° = 108°$. So each base angles angle is $54°$.

From the diagram, these base angles are each half of an interior angle, so this gives another way of working out interior angles, too.

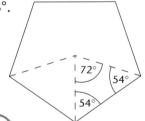

EXAMPLE 6

Find the interior angle of a regular octagon.

For a regular octagon, the exterior angle $= 360° \div 8 = 45°$.

So the interior angle $= 180° - 45° = 135°$.

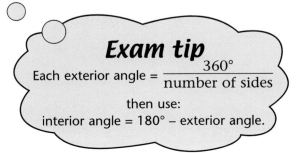

Exam tip

Each exterior angle $= \dfrac{360°}{\text{number of sides}}$

then use:

interior angle $= 180° - $ exterior angle.

Exam tip

A pentagon has 5 sides, a hexagon 6, a heptagon 7, an octagon 8, a nonagon 9 and a decagon 10.

EXERCISE 3.4A

1 Three of the exterior angles of a quadrilateral are 85°, 55° and 90°. Find the size of the other exterior angle.

2 Four of the exterior angles of a pentagon are 73°, 60°, 81° and 90°. Find the size of the other exterior angle.

3 Five of the exterior angles of a hexagon are 44°, 58°, 63° 40° and 82°. Find the size of the other exterior angle.

4 Find the sizes of the interior angles of the pentagon in question 2.

5 Find the sizes of the interior angles of the hexagon in question 3.

6 A regular polygon has eight sides. Find the sizes of its exterior and interior angles.

7 Find the interior angle of a regular nonagon (9 sides).

8 A regular polygon has an exterior angle of 20°. How many sides does it have?

9 What is the sum of the interior angle of:

   a) a heptagon
   b) a an octagon?

10 Six of the angles of a heptagon are 127°, 142°, 131°, 104°, 163° and 128°. Calculate the size of the remaining angle.

EXERCISE 3.4B

1 Three of the exterior angles of a quadrilateral are 114°, 60° and 71°. Find the size of the other exterior angle.

2 Four of the exterior angles of a pentagon are 62°, 47°, 88° and 79°. Find the size of the other exterior angle.

3 Four of the exterior angles of a hexagon are 61°, 47°, 93° and 35°. Find the size of the other exterior angles, given that they are equal.

4 Find the sizes of the interior angles of the pentagon in question 2.

Exercise 3.4B cont'd

5 Find the sizes of the interior angles of the hexagon in question 3.

6 A regular polygon has 12 sides. Find the sizes of its exterior and interior angles.

7 Find the interior angle of a regular 18-sided polygon.

8 A regular polygon has an exterior angle of 18°. How many sides does it have?

9 What is the sum of the interior angles of:

a) a nonagon

b) a dodecagon (12 sides)?

10 A polygon has 11 sides. Ten of its interior angles add up to 1500°. Find the size of the remaining angle.

Key ideas

Angles on a straight line add up to 180°.

Angles round a point add up to 360°.

When two straight lines cross, the opposite angles are equal.

4 Probability

You should already know

- that probabilities are expressed as fractions, decimals or percentages
- that all probabilities lie on a scale of 0–1 (0–100%)
- how to find a probability from a set of equally likely outcomes:
  for example, the probability of throwing a six with a fair die is $\frac{1}{6}$ the probability of choosing a king from a pack of ordinary playing cards is $\frac{4}{52} = \frac{1}{13}$.
- how to subtract fractions and decimals from 1 for example, $1 - \frac{3}{7} = \frac{4}{7}$ and $1 - 0.4 = 0.6$.

Probability of an outcome not happening

When a fair die is thrown, this is called an event.
In this example, there are six equally likely scores or outcomes: 1, 2, 3, 4, 5 and 6.

So the probability of throwing a six is $\frac{1}{6}$.

The probability of *not* throwing a 6 is the same as the probability of throwing a 1 *or* 2 *or* 3 *or* 4 *or* 5, which is $\frac{5}{6}$.
Now $\frac{5}{6} = 1 - \frac{1}{6}$, so this gives the method of finding the probability that an outcome does not occur.

> If the probability of an outcome happening is p, the probability of the outcome **not** happening = $1 - p$.

Instead of writing 'the probability that an outcome happens', you can use the shorter P(outcome).

So in the above example: P(getting a six) = $\frac{1}{6}$
P(not getting a six) = $\frac{5}{6}$

EXAMPLE 1

The probability that Nayim's school bus is late is $\frac{1}{8}$.
What is the probability that Nayim's bus is not late?

Probability that Nayim's bus is not late
= 1 − (probability that Nayim's bus is late)
= $1 - \frac{1}{8}$
= $\frac{7}{8}$

EXAMPLE 2

The probability that it will rain tomorrow is 0·3.
What is the probability that it will not rain tomorrow?

Probability that it will not rain tomorrow

= 1 − (probability that it will rain tomorrow)

= 1 − 0·3

= 0·7.

Exam tip

Never write probabilities as '1 in 4', '1 to 4', '1 out of 4' or '1 : 4'. You will lose marks if you do. Probabilities should be written as fractions or decimals, for example $\frac{1}{4}$ or 0·25. Fractions are usually preferable as they are exact. If, however, the probabilities in the question are given as decimals, then it is usually better to use decimals. Percentages are acceptable but since they usually involve extra work there is little point in using them.

EXERCISE 4.1A

1 The probability of United winning their next game is 0·8. What is the probability of United not winning their next game?

2 The probability that I will have cereal for breakfast is $\frac{2}{7}$. What is the probability that I will not have cereal for breakfast?

3 The picture shows a fair spinner.

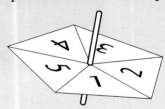

a) What is the probability of getting an even number with one spin?

b) What is the probability of getting an odd number with one spin?

4 There are 13 diamonds in a pack of 52 playing cards. I choose a card at random.

a) What is the probability that I choose a diamond?

b) What is the probability that I do not choose a diamond?

5 In a multiple-choice test paper there are five possible answers to each question, only one of which is right. If Obaid does not know an answer to a question, he guesses. Obaid guessed the answer to question 15.

a) What is the probability that Obaid got question 15 right?

b) What is the probability that Obaid got question 15 wrong?

Exercise 4.1A cont'd

6 The probability that the first ball drawn in the lottery draw is white is $\frac{9}{49}$. What is the probability that it is not white.

7 The probability that Peter goes out on a Friday night is 0·995. What is the probability that Peter stays in on a Friday night?

EXERCISE 4.1B

1 The probability that Elma will pass her next Maths examination is 0·85. What is the probability that she will fail the examination?

2 Based on past experience, the probability that Kevin's school bus will be late tomorrow is 0·23. What is the probability that it is on time or early?

3 The probability that Jane will do the washing up tonight is $\frac{3}{8}$. What is the probability that she will not do the washing up?

4 The probability that Barry will get a motorcycle for his sixteenth birthday is 0·001. What is the probability that he will not get a motorcycle?

5 The probability that Zebedee will go to bed on time is $\frac{1}{25}$. What is the probability that he will not go to bed on time?

6 The probability that Liam will watch City on Saturday is 0·98. What is the probability that he will not watch City?

7 The probability that Umair will choose to study Geography at GCSE is $\frac{2}{7}$. What is the probability that he will not choose Geography?

Mutually exclusive outcomes

Mutually exclusive outcomes are outcomes which cannot happen together. For example if you toss a coin once then the outcomes 'the coin comes down heads' and 'the coin comes down tails' are mutually exclusive, since it is impossible for the coin to come down both heads and tails.

If outcomes are mutually exclusive and cover all the possibilities then the probabilities of those events must add up to 1.

So in the above example:

P(heads) + P(tails) = 1.

This would be true even if the coin was not fair but was biased towards heads.

If P(heads) = 0·6 then P(tails) = 0·4 and P(heads) + P(tails) = 1 since you cannot throw both a head and a tail on the same coin on the same toss, and no other outcome is possible.

EXAMPLE 3

The probability that City win their next game is 0·5. The probability that they lose the game is 0·2. What is the probability that the game will be drawn?

The outcomes 'win', 'lose' and 'draw' are mutually exclusive, as it is impossible for any two or all of them to happen together.

Also no outcomes, other than 'win', 'lose' and 'draw', are possible.

Therefore P(win) + P(lose) + P(draw) = 1.

P(draw) = 1 − P(win) − P(lose) = 1 − 0·5 − 0·2 = 0·3.

EXERCISE 4.2A

1   In a game of tennis you can only win or lose. A draw is not possible. Qasim plays Robin at tennis. The probability that Qasim wins is 0·7. What is the probability that Robin wins?

2   A bag contains red, white and blue balls. I pick one ball out of the bag at random. The probability that I pick a red one is $\frac{1}{12}$. The probability that I pick a white one is $\frac{7}{12}$. What is the probability that I pick a blue one?

3   For my next holiday I will go to Spain or France or the USA. The probability that I will go to Spain is $\frac{7}{20}$. The probability that I will go to France is $\frac{11}{20}$. What is the probability that I will go to the USA?

4   The probability that the school team will win their next match is 0·4. The probability that they will lose the match is 0·5. What is the probability that they will draw the match?

5   Sally has only grey, navy and black skirts in her wardrobe. She is choosing a skirt to wear to the cinema. The probability that she chooses a grey one is 0·2. The probability that she chooses a navy one is 0·15. What is the probability that she chooses a black one?

Exercise 4.2A cont'd

6 Max travels to school by bus, car or cycle or he walks. The probability that he travels by bus is $\frac{1}{11}$. The probability that he travels by car is $\frac{3}{11}$. The probability that he cycles is $\frac{2}{11}$. What is the probability that he walks?

7 Tariq hears on the weather forecast that the probability that it will rain tomorrow is 0·6. He says that this means that the probability that the sun will shine is 0·4. Give two reasons why he may not be correct.

EXERCISE 4.2B

1 A set of traffic lights may be on red, red and amber, amber or green. The probability that they are on red is 0·5. The probability that they are on red and amber is 0·05 and the probability that they are on amber is 0·05. What is the probability that the lights are on green?

2 Alex is choosing his GCSE options. In one pool, he can choose History or Geography or Business Studies. The probability that he chooses History is 0·3. The probability that he chooses Geography is 0·45. What is the probability that he chooses Business Studies?

3 The coach is selecting the netball team. She has three goal attacks to choose from, Raisa, Kimberley and Melanie. The probability that she chooses Raisa is $\frac{3}{8}$. The probability that she chooses Kimberley is $\frac{5}{8}$. What is the probability that she chooses Melanie?

4 There are four breakfast cereals in the cupboard: muesli, cornflakes, Weetycrisps and Frosties. Kim has decided to have cereal for breakfast. The probability that she has muesli is 0·05. The probability that she has Weetycrisps is 0·4. The probability that she has Frosties is 0·2. What is the probability that she will have cornflakes?

5 Greg is choosing his main course for his school dinner. There are three choices: burger and chips, tuna salad and tagliatelli. The probability that he chooses tuna salad is $\frac{1}{12}$. The probability that he chooses tagliatelli is $\frac{5}{12}$. What is the probability that he chooses burger and chips?

Exercise 4.2B cont'd

6   There are red, yellow and blue beads in a bag. The probability of choosing a red one is $\frac{1}{3}$. The probability of choosing a yellow one is $\frac{1}{4}$. What is the probability of choosing a blue one?

7   Outcomes A, B and C are mutually exclusive and one of them must happen. If P(A) = 0·47 and P(B) = 0·31, what is P(C)?

8   The probability that Rovers will win their next game is 0·4. Geri says this means that the probability they will lose is 0·6. Why is she almost certainly wrong?

Key ideas

- If the probability of an outcome happening is p, then the probability of the outcome not happening = $1 - p$.

- If two outcomes A and B cannot occur together they are mutually exclusive.

- If two outcomes A and B are mutually exclusive and cover all possible outcomes, then P(A) + P(B) = 1. Similarly, if events A, B and C are mutually exclusive and cover all possible outcomes then: P(A) + P(B) + P(C) = 1 and so on.

Revision exercise

1 Work these out using a calculator.
  a) $(^-2{\cdot}3 \times \ ^-4{\cdot}6) + (^-3{\cdot}9 \times 2{\cdot}1)$
  b) $4{\cdot}2 + 6{\cdot}3 - 8{\cdot}4 - 7{\cdot}9 + 1{\cdot}3 + 3{\cdot}1 - 5{\cdot}2 - 6{\cdot}3$
  c) $\dfrac{^-3{\cdot}2 \times 2{\cdot}3 + 7{\cdot}9 \times \ ^-2{\cdot}4}{^-8{\cdot}4}$

2 Work these out using a calculator.
  a) $^-2{\cdot}73 + 12{\cdot}6 - 11{\cdot}91 + 13{\cdot}2$
  b) $(^-4{\cdot}5 \times 8{\cdot}3) + (6{\cdot}1 \times \ ^-4{\cdot}3)$
  c) $\dfrac{^-4{\cdot}7 + 3{\cdot}6}{^-7{\cdot}5}$

3 Work these out using a calculator.
  a) 3^4
  b) 4^4
  c) $12{\cdot}3^3 - 2{\cdot}6^3$

4 Work these out using a calculator.
  a) $\sqrt{5^2 + 8^2}$
  b) $\dfrac{7{\cdot}92 \times 1{\cdot}71}{4{\cdot}2 + 3{\cdot}6}$
  c) $(4{\cdot}1 - 3 \times 2{\cdot}6)^3$

5 Expand these brackets.
  a) $4(a - 3)$
  b) $5(3 + x)$
  c) $7(a + 2)$
  d) $3(3 + 3x)$
  e) $5(9x + 7)$

6 Factorise these fully.
  a) $12a - 6$
  b) $15x^2 - x$
  c) $4a^2 + a$
  d) $a^2 + 4a$
  e) $3y - 7y^2$

7 An isosceles triangle has an angle of 50°. Find the sizes of the other angles in the triangle. There are two possible sets of answers – find both sets.

8 Three angles of a quadrilateral are 67°, 122° and 94°. Find the size of the other angle.

9 Find the sizes of the lettered angles, giving your reasons.

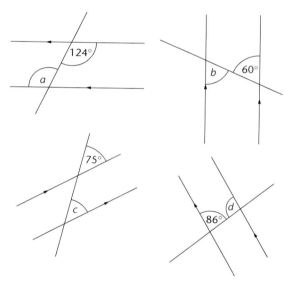

10 Four of the exterior angles of a pentagon are 84°, 67°, 55° and 96°. Find the size of the other exterior angles.

11 The probability that Waseem will get all his Maths homework correct is 0·65. What is the probability that he will not get it all correct?

12 The probability that James will play computer games tonight is $\frac{7}{9}$. What is the probability that he will not play computer games tonight?

13 In a game of badminton one player must win. A draw is not possible. Graeme plays Chris in three games of badminton. The probability that Graeme wins all three games is $\frac{1}{8}$. What is the probability that Chris wins at least one game?

14 There are green, black and white balls in a bag. One ball is selected at random. The outcomes are denoted by G, B and W.

P(G) = 0·43, P(B) = 0·28. Find P(W).

15

> ## MENU
> Main course
> Sausage & Chips
> Ham Salad
> Vegetable Lasagne

Melanie is choosing one main course from a choice of three on the canteen menu. The probability that she chooses sausage and chips is 0·1. The probability that she chooses ham salad is 0·6. What is the probability that she chooses vegetable lasagne?

5 Percentages

You should already know

- how to write and read fractions, decimals and percentages
- how to multiply and divide numbers and decimals by 100.

You should also know that percentage means 'out of 100', for example:

'75% of cat owners surveyed said their cats preferred fish flavoured treats'

75% means that 75 out of every 100 people surveyed said their cats preferred fish flavoured treats.

'40% of people asked said they drank coffee at breakfast'

40% means that 40 people out of every 100 people drank coffee at breakfast.

Fractions to decimals to percentages

Seven out of 10 households buy their milk at supermarkets.

What is this as a percentage?

You can work this out using equivalent fractions:

$\frac{7}{10} = \frac{70}{100} = 70\%$ multiplying the numerator and the denominator by 10;
or by changing the fraction to a decimal:

$\frac{7}{10} = 0 \cdot 7 = 70\%$.

To change a fraction into a decimal remember to divide the numerator by the denominator – see example 1.

EXAMPLE 1

Convert $\frac{3}{5}$ to **a)** a decimal **b)** a percentage.

a) $\frac{3}{5} = 3 \div 5 = 0 \cdot 6$ (This can be done on a calculator, or as $5\overline{)3 \cdot 0}$ with $0 \cdot 6$ above).

b) $0 \cdot 6 \times 100 = 60\%$.

NB It is possible to go straight from the fractions to the percentage using multiplications of fractions.

$\frac{3}{5} \times 100 = \frac{3}{5} \times \frac{100}{1} = \frac{300}{5} = 60\%$.

EXERCISE 5.1A

1 Copy the following table and complete it.

Fraction $\dfrac{a}{b}$	Decimal $a \div b$	Percentage = decimal × 100
$\frac{7}{10}$		
$\frac{2}{5}$		
$\frac{3}{4}$		
$\frac{1}{3}$		
$\frac{2}{3}$		

2 Change the following decimals to percentages.

a) 0·37　　**b)** 0·83　　**c)** 0·08

d) 0·345　　**e)** 1·25.

3 Change the following fractions to decimals.

a) $\frac{1}{100}$　　**b)** $\frac{17}{100}$　　**c)** $\frac{2}{50}$　　**d)** $\frac{8}{5}$.

4 On Wednesday 0·23 of the population watched Westenders. What percentage is this?

5 At a matinée, $\frac{4}{5}$ of the audience at the pantomime were children. What percentage is this?

6 In a survey, $\frac{3}{10}$ of children liked cheese and onion crisps. What percentage is this?

Exam tip

For the non-calculator paper it is worth learning some basic equivalents.

$\frac{1}{2} = 0\cdot5 = 50\%$　　$\frac{1}{3} = 0\cdot333\ldots = 33\cdot3\ldots\%$　　$\frac{1}{5} = 0\cdot2 = 20\%$

$\frac{1}{4} = 0\cdot25 = 25\%$　　(33% to the nearest 1%)　　$\frac{2}{5} = 0\cdot4 = 40\%$

$\frac{3}{4} = 0\cdot75 = 75\%$　　$\frac{2}{3} = 0\cdot666\ldots = 66\cdot6\ldots\%$　　$\frac{3}{5} = 0\cdot6 = 60\%$

　　　　　　　　(67% to the nearest 1%)　　$\frac{4}{5} = 0\cdot8 = 80\%$

Notice the rounding of $\frac{1}{3}$ and $\frac{2}{3}$. It is a common error to assume $\frac{1}{3} = 0\cdot3 = 30\%$ and $\frac{2}{3} = 0\cdot66 = 66\%$ or even $\frac{2}{3} = 0\cdot6 = 60\%$.

EXERCISE 5.1B

1 Change the following fractions to decimals.

a) $\frac{1}{8}$ **b)** $\frac{3}{8}$ **c)** $\frac{3}{20}$

d) $\frac{17}{40}$ **e)** $\frac{5}{16}$

2 Change the decimals you found in question 1 to percentages.

3 Change the following fractions into percentages. Give your answers correct to one decimal place.

a) $\frac{1}{6}$ **b)** $\frac{5}{6}$ **c)** $\frac{5}{12}$

d) $\frac{1}{15}$ **e)** $\frac{3}{70}$

4 A kilometre is $\frac{5}{8}$ of a mile. What percentage is this?

5 The winning candidate in an election gained $\frac{7}{12}$ of the votes. What percentage is this? Give your answer correct to the nearest 1%.

6 Nicola spends $\frac{3}{7}$ of her pocket money on sweets and drinks. What percentage is this? Give your answer correct to the nearest 1%.

7 Imran has a part time job. He saves $\frac{2}{9}$ of his wages. What percentage is this? Give your answer correct to the nearest 1%.

Expressing one quantity as a percentage of another

To express one quantity as a percentage of another, start by writing the first quantity as a fraction of the second.

Then use the methods already described to change the fraction to a percentage.

Exam tip

To do this both quantities must be in the same units.

EXAMPLE 2

Express 4 as a percentage of 5.

First write 4 as a fraction of 5: $\frac{4}{5}$.

Then change $\frac{4}{5}$ to a percentage: $\frac{4}{5} = 0.8 = 80\%$.

EXAMPLE 3

Express £5 as a percentage of £30.

$\frac{5}{30} = 0.167 = 16.7\%$ to one decimal place.

EXAMPLE 4

Express 70 cm as a percentage of 2·3 m.

First change 2·3 m to centimetres.

2·3 m = 230 cm

$\frac{70}{230} = 0.304 = 30.4\%$ to 1 decimal place.

or

change 70 cm to metres: 70 cm = 0·7 m.

$\frac{0.7}{2.3} = 0.304 = 30.4\%$ to one decimal place.

EXERCISE 5.2A

1 In each case, express the first quantity as a percentage of the second.

	First quantity	Second quantity
a)	16	100
b)	12	50
c)	1 m	4 m
d)	£3	£10
e)	73p	£1
f)	8p	£1
g)	£1·80	£2
h)	40 cm	2 m
i)	50p	£10
j)	£2·60	£2.

2 An article in a shop costs £5, then the shopkeeper increases the price by £1. What percentage increase is this?

3 A sailor has a rope 50 m long. He cuts off 12 m. What percentage of the rope has he cut off?

4 In a sales promotion a company offers an extra 150 ml of cola free. The bottles usually hold 1 litre. What percentage is 150 ml of 1 litre?

5 After thermal insulation was installed, Saima's central heating bill was reduced by £140 per year. If the bill was previously £500, what is the percentage reduction?

EXERCISE 5.2B

1 In each case, express the first quantity as a percentage of the second. Where appropriate give your answer to the nearest 1%.

	First quantity	Second quantity
a)	11	16
b)	9	72
c)	1 m	7 m
d)	£3	£18
e)	73p	£3
f)	17p	£2·50
g)	£2·60	£7
h)	40 cm	2·6 m
i)	£3·72	£12·96
j)	£2·85	£2·47.

2 A shopkeeper reduces the cost of an item by £2. If it originally cost £11, what is the percentage reduction? Give your answer to one decimal place.

3 A spring is originally 60 cm long. It is stretched by 17 cm. By what percentage is it stretched? Give your answer to one decimal place.

4 A school has 857 pupils and 70 of them go on a school holiday. What percentage is this? Give your answer to one decimal place.

5 A mathematics test has 45 questions. Abigail got 42 right. What percentage did Abigail get right? Give your answer to the nearest 1%.

Percentage increase and decrease

In this section you will solve problems, using methods you have learnt previously.

EXAMPLE 5

A shopkeeper buys an article for £25 and sells it for £30. What percentage profit is this?

Profit = £30 − £25 = £5.

Percentage profit = $\frac{5}{25} \times 100 = 20\%$.

Exam tip

Percentage increases and decreases are workout out as percentages of the original amount, not the new amount. Percentage profit or loss is worked out as a percentage of the cost price, not the selling price.

EXAMPLE 6

A car drops in value from £7995 to £7000 in a year. What percentage decrease is this?

Decrease in value = £7995 − £7000 = £995

Percentage decrease = $\frac{995}{7995} \times 100 = 12.4\%$ to one decimal place.

Exam tip

Examples 7 and 8 can be done by a shorter method. This method will make calculations much easier when doing repeated calculations and, in later work, when working back from an answer. The method is shown in Examples 9 and 10. It is particularly useful on the calculator paper.

EXAMPLE 7

An engineer receives a 3% increase in her annual salary. If she earned £24 000 before the increase, what is her new salary?

Increase = $24\,000 \times \frac{3}{100} = 24\,000 \times 0.03 = 720$.

New salary = £24 000 + £720 = £24 720.

EXAMPLE 9

Increase £24 000 by 3%.

You need to calculate 3% of £24 000 and add it on to the original £24 000.

You need 3% of 24 000 + 100% of 24 000.

This is the same as 103% of £24 000.

The calculation now becomes

£24 000 × $\frac{103}{100}$ = £24 000 × 1·03 = £24 720.

Using this method, you carry out the percentage calculation and the addition in one step.

EXAMPLE 8

In a sale all prices are reduced by 15%. Find the new price of an article previously priced at £17·60.

SALE
15% off
everything!

Reduction = $17·60 \times \frac{15}{100}$

= 17·60 × 0·15 = 2·64.

New price = £17·60 − £2·64

= £14·96.

EXAMPLE 10

Reduce £17·60 by 15%.

You need to find 15% of £17·60 and subtract the answer from £17·60.

You need 100% of £17·60 − 15% of £17·60.

This is the same as 85% of £17·60 (since 100% − 15% = 85%).

The calculation now becomes £17·60 × $\frac{85}{100}$ = £17·60 × 0·85 = £14·96.

Using this method, you carry out the percentage calculation and the subtraction in one step.

EXAMPLE 11

Due to inflation, prices increase by 5% per year. An item costs £12 now. What will it cost in 2 years' time?

In 1 year the price will be
£12 × 1·05 = £12·60.

In 2 years the price will be
£12·60 × 1·05 = £13·23.

Alternatively, this repeated calculation could be worked out as:
£12·60 × 1·05 × 1·05 = £13·23.

EXAMPLE 12

Each year a car loses value by 12% of its value at the beginning of the year. If its starting value was £9000 find its value after 3 years.

100% − 12% = 88%.

So after 1 year the value = £9000 × 0·88 = £7920.

after 2 years the value = £7920 × 0·88 = £6969·60.

after 3 years the value = £6969·60 × 0·88 = £6133·25 (to the nearest penny).

Alternatively, this repeated calculation could be worked out as:

£9000 × 0·88 × 0·88 × 0·88 = £6133·25.

EXERCISE 5.3A

1 A shopkeeper buys an article for £10 and sells it for £13. What percentage profit does she make?

2 David earned £3 per hour. When the National Minimum Wage was introduced his pay increased to £3·60 per hour. What was the percentage increase in David's pay?

3 A season ticket for Newtown Rovers normally costs £500. If it is bought before 1 June it costs £240. What percentage reduction is this?

4 Lee bought a CD for £12·50. A year later he sold it for £5. What percentage of the value did he lose?

Exercise 5.3A cont'd

5   Ghalib's gas bill is £240 before VAT is added on. What is the bill after VAT at 5% is added on?

6   Kate's diet has led to her losing 15% of her weight. If she weighed 80 kg before starting her diet, what does she weigh now?

7   Train fares went up by 8%. What is the new price of a ticket which previously cost £3?

8   In a sale, prices are reduced by 20%. What is the sale price, if the original price was £13?

Exam tip

V.A.T. means value added tax. This tax is added to the basic price for goods and services.

EXERCISE 5.3B

1   A shopkeeper buys an article for £22 and sells it for £27. What is his percentage profit? Give your answer to the nearest 1%.

2   A rail company reduces the time for a journey from 55 minutes to 48 minutes. What percentage reduction is this? Give your answer to the nearest 1%.

3   Claire receives a pay increase from £135 per week to £145 per week. What percentage rise is this? Give your answer to one decimal place.

4   In a sale the price of a dress is reduced from £60 to £39. What percentage reduction is this?

5   A1 Electrics buys washing machines for £255 and sells them for £310. Bob's Budget Bargains buys washing machines for £270 and sells them for £330. Which company makes the greater percentage profit? You must show all your working.

6   To test the strength of a piece of wire it is stretched by 12% of its original length. If it was originally 1·5 m long, what will be its length after stretching?

7   Rushna pays 6% of her pay into a pension fund. If she earns £185 per week, what will her pay be after taking off her pension payments?

8   A mini tower stereo system costs £280 before V.A.T. is added on. What will it cost after V.A.T. at 17·5% is added on?

Exercise 5.3B cont'd

9 Mick puts £220 into a savings account. Each year the savings earn interest at 7% of the amount in the account at the start of the year. What will his savings be worth after 2 years? Give your answer to the nearest penny.

10 Each year a car loses value by 12% of its value at the start of the year. If it is worth £9000 when it was new, what will it be worth after 3 years?

Key Idea

To change a fraction into a decimal, divide the top of the fraction by the bottom. $\frac{a}{b} = a \div b$.

To change a decimal into a percentage, multiply by 100.

To find one quantity as a percentage of another, write the first quantity as a fraction of the second. Then change to a percentage as above. (Both quantities must be in the same units).

To increase a quantity by, for example, 5%, a quick way is to multiply by 1·05.

To reduce a quantity by, for example, 12%, a quick way is to multiply by 0·88 (as 100 − 12 = 88).

6 Ratio

You should already know

- how to simplify fractions
- how to find common factors.

A reminder

When something is divided into a number of parts, ratios are used to compare the sizes of those parts.

For example:

if 10 sweets were divided between two people so one person got six sweets and the other four sweets, the sweets have been divided in the ratio $6:4$;

if £30 were divided between two people in the ratio $3:2$ one person would receive three parts of the £30 and the other two parts of the £30.

- As with fractions ratios can be simplified and should be if possible.
  As a fraction $\frac{6}{4} = \frac{3}{2}$; as a ratio $6:4 = 3:2$.
- If units are used in ratios the units must be the same for both numbers when simplifying.
  Thus 40 minutes : 1 hour should be 40 minutes : 60 minutes = $2:3$
  and $1\,\text{kg} : 300\,\text{g} = 100\,\text{g} : 300\,\text{g} = 10:3$.

ACTIVITY 1

Remember when simplifying ratios look for common factors. Thus $40:60$ will simplify to $2:3$ by dividing both numbers by 20. You could have arrived at $2:3$ by dividing by different factors. For example:
$40:60 = 4:6 = 2:3$ (here dividing by 10 and then 2);
or $40:60 = 20:30 = 10:15 = 2:3$ (by dividing by 2, then 2 and then 5).

Find as many ways as you can to simplify the following ratios:

a) $20:50$ **b)** $16:20$ **c)** $14:56$
d) $9:27$ **e)** $6:48$.

EXERCISE 6.1A

1 Write each ratio in its simplest form.

   a) $6:3$ **b)** $25:75$ **c)** $30:6$
   d) $10:15$ **e)** $7:35$.

2 Write each ratio in its simplest form:

   a) 20 minutes : 1 hour
   b) $50\,\text{g} : 1\text{kg}$ **c)** $300\,\text{ml} : 1$ litre
   d) $2\,\text{kg} : 600\,\text{g}$
   e) 2 minutes : 30 seconds

EXERCISE 6.1B

1 Write each ratio in its simplest form.

  a) 15:12   **b)** 24:8   **c)** 4:48

  d) 3:27   **e)** 9:81.

2 Write each ratio in its simplest form.

  a) 30p:£1   **b)** 2 m:50 cm

  c) 45 seconds:2 minutes

  d) 5 kg:750 g

  e) 500 m:3 km.

Using ratios

EXAMPLE 1

Two families are going to Alton Towers for the day.

There are five people in the Smith family and four people in the Jones family.

They rent a minibus for the day. It costs £90.

How much does each family pay?

There are nine people altogether, so each person will pay $\frac{1}{9}$th of £90 = £10. The Smith family pays 5 × £10 = £50; the Jones family pays 4 × £10 = £40.

EXAMPLE 2

Asif and Jane work together as partners in a shop. Asif works 2 days and Jane works 4 days. They share the weekly profit in the ratio of the days worked.

If the profit in one week was £1800 how much did they each get?

They are 6 days altogether so for 1 day they would receive $\frac{£1800}{6}$ = £300.

So Asif receives 2 × £300 = £600 and Jane receives 4 × £300 = £1200.

EXERCISE 6.2A

1 A 600 g bar of brass is made using the metals copper and zinc in the ratio 2:1. How much of each metal is used?

2 Tom and Tanya share an £80 bonus in the ratio 3:5. How much does each receive?

3 There are 120 employees in a factory. The ratio of males to females is 5:7. How many employees are female?

4 In a class of 36 children the ratio of boys to girls is 5:4. How many boys are there?

EXERCISE 6.2B

1 Share £750 in the ratio 3 : 1.

2 Fertiliser needs mixing with water in the ratio 20 ml to 1 litre. How much fertiliser would you need with 4 litres of water?

3 In a university lecture there are 60 female students and 12 male students. What is the ratio of males to females?

4 For a football match tickets are allocated between the 'home' club and the 'away' club in the ratio 6 : 1. How many tickets does the home club receive if the total number of tickets available is 35 000?

Using ratios continued

When you are mixing quantities it is often important to keep the amounts in the same proportion.

For example, if you are mixing black paint and white paint to make a certain shade of grey, it will be important to keep the proportions of black and white paint the same.

If the colour you want is obtained by mixing 2 litres of black paint with 1 litre of white paint, then you will need 4 litres of black paint if you use 2 litres of white paint.

The ratio of black paint to white paint is 2 parts to 1 part.

EXAMPLE 3

To mix the same shade of grey paint as above, how much white paint will you need to mix with 8 litres of black paint?

To keep track of ratio questions, it is often helpful to form a table.

Black paint   **White paint**

The multiplier for black paint is 4, so you must use the same multiplier for white paint:

$1 \times 4 = 4$

So you need 4 litres of white paint.

Chapter 6 *Ratio*

EXAMPLE 4

To make pink paint, red paint is added to white paint in the ratio 3 parts red to 2 parts white.

a) How much white paint should be mixed with 6 litres of red paint?

b) How much red paint should be mixed with 10 litres of white paint?

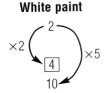

Red paint   **White paint**

For **a)** the multiplier is 2 so you need $2 \times 2 = 4$ litres of white paint.

For **b)** the multiplier is 5 so you need $3 \times 5 = 15$ litres of red paint.

If the multiplier is not immediately obvious you may need to use division to find it.

EXAMPLE 5

To make light grey paint, Rosie mixes black and white paint in the ratio 1 part black to 5 parts white. How much black paint will she need to mix with 7 litres of white paint?

Black paint   **White paint**

Here the multiplier is $7 \div 5 = 1 \cdot 4$.

So she needs $1 \times 1 \cdot 4 = 1 \cdot 4$ litres of black paint.

EXERCISE 6.3A

1 Sanjay is mixing light pink paint. He mixes red paint to white paint in the ratio 1 part red to 3 parts white.

    a) How much white paint should he mix with 3 litres of red paint?

b) How much red paint should he mix with 12 litres of white paint?

2 Michelle is making mortar. To do this she mixes sand and cement in the ratio 5 parts sand to 1 part cement. She measures the quantities in bags.

Exercise 6.3A cont'd

a) How many bags of sand should she mix with 2 bags of cement?

b) How many bags of cement should she mix with 20 bags of sand?

3 Julia is making a cake. To do this she starts by mixing flour and fat in the ratio 8 parts flour to 3 parts fat.

a) How much fat should she mix with 800 grams of flour?

b) How much flour should she mix with 60 grams of fat?

4 Marco is making jam. To do this he mixes fruit and sugar in the ratio 2 parts fruit to 3 parts sugar.

a) How much sugar should he mix with 8 kg of fruit?

b) How much fruit should he mix with 15 kg of sugar?

5 A chemist is making a solution of a chemical. To do this he mixes 1 part of the chemical to 20 parts water.

a) How much water should he mix with 10 ml of the chemical?

b) How much chemical should he mix with 2 litres of water? (1 litre = 1000 ml)

6 The same chemist is now making a weaker solution of the chemical. To do this he mixes 1 part of the chemical to 50 parts of water.

a) How much water should he mix with 10 ml of the chemical?

b) How much chemical should he mix with 2 litres of water? (1 litre = 1000 ml)

EXERCISE 6.3B

1 Graham is making pastry. To do this he starts by mixing flour and fat in the ratio 8 parts flour to 3 parts fat.

a) How much fat should he mix with 320 grams of flour?

b) How much flour should he mix with 150 grams of fat?

2 Kate is mixing dark pink paint. She mixes red and white paint in the ratio 4 parts red to 3 parts white.

a) How much white paint should she mix with 12 litres of red paint?

b) How much red paint should she mix with 15 litres of white paint?

Exercise 6.3B cont'd

3 Sarah and Heather share a flat. They agree to spend time doing the cleaning in the ratio 2 parts (Sarah) to 3 parts (Heather).

a) If Sarah spends 5 hours cleaning, how long should Heather spend?

b) If Heather spends 4 hours cleaning, how long should Sarah spend?

4 Regulations for school trips say that there must be at least two adults for every 30 pupils.

a) How many adults should there be for 75 pupils?

b) How many pupils can go on a trip if there are seven adults available?

c) How many adults should there be for 200 pupils?

5 Rashid fills his flower tubs with a mixture of 4 parts soil to 3 parts compost.

a) How much soil should he mix with 45 litres of compost?

b) How much compost should he mix with 68 litres of soil?

6 Orange squash needs to be mixed in the ratio 1 part concentrate to 6 parts water.

a) How much water should be mixed with 80 ml of concentrate?

b) How much concentrate should be mixed with 2 litres of water?
(1 litre = 1000 ml)

Sharing in a given ratio

In the previous section you saw how to mix grey paint if you know how much black or white paint to use. In this section you will find out how much black and white paint to use, to make up the amount of grey paint you need.

In Example 3, black and white paint were mixed in the ratio 2 parts black to 1 part white. How much of each colour paint will you need, to mix 12 litres of grey paint?

Two litres of black paint and 1 litre of white paint will make 3 litres of grey paint.

You can solve this problem in a similar way, using a table with three columns.

Black paint	White paint	Mixture (grey)
2	1	3
×4	×4	×4
8	4	12

The multiplier is 4.

So you need: 2 × 4 = 8 litres of black paint and 1 × 4 = 4 litres of white paint.

Now check that the answers add up to 12.

Exam tip

When sharing in a given ratio you always need to add the numbers in the ratio together.

Exam tip

Work out the shares separately and then check that they add up to the right amount.

EXAMPLE 6

A recipe for white sauce says, 'Mix butter, flour and milk in the ratio 1 part butter to 1 part flour to 10 parts milk.'
How much of each ingredient is needed to make 240 grams of white sauce?

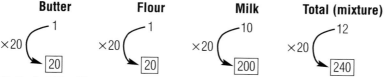

| Butter | Flour | Milk | Total (mixture) |

The multiplier is 240 ÷ 12 = 20.
So you need 1 × 20 = 20 g each of butter and flour and 10 × 20 = 200 g of milk.
Check: 20 + 20 + 200 = 240.

EXERCISE 6.4A

1 Share £20 between Charles and David in the ratio 2 parts to 3 parts.

2 A 20% solution is made by mixing 1 part of a chemical to 4 parts of water. How much chemical will there be in 100 ml of the solution?

3 Light pink paint is made by mixing red and white paint in the ratio 1 part red to 3 parts white. How much of each colour is needed to make 16 litres of light pink paint?

4 Dark grey paint is made by mixing black and white paint in the ratio 5 parts black to 2 parts white. How much of each colour is needed to make 35 litres of dark grey paint?

5 Asif is mixing mortar by mixing sand and cement in the ratio 5 parts sand to 1 part cement. How much sand is needed to make 36 kg of the mixture?

Exercise 6.4A cont'd

6 Orange squash needs to be mixed in the ratio 1 part concentrate to 6 parts water. How much concentrate is needed to make 3·5 litres of orange squash?

7 Share £18 among Eileen, Fiona and George in the ratio 4 parts to 2 parts to 3 parts.

EXERCISE 6.4B

1 Share £28 between Hamid and Ian in the ratio 5 parts to 3 parts.

2 In the audience for a pantomime, the ratio of adults to children is 1 parts to 5 parts. If there are 630 in the audience, how many children are there?

3 Jane is cutting a rope into two pieces in the ratio 7 parts to 3 parts. If the length of the rope is 3 m, how long will each of the two pieces be?

4 Light grey paint is made by mixing black paint and white paint in the ratio 4 parts black to 11 parts white. How much of each colour will be needed to make 180 litres of light grey paint?

5 The number of members in a parliament is shared among three parties in the ratio 4 parts to 3 parts to 2 parts. If there are a total of 657 members, how many of each party are there?

6 To get home Michael runs and walks. The distance he runs and walks are in the ratio 2 parts to 3 parts. If he works 2 km from home, how far does he run?

7 Three children share £50 in the ratio of their ages. Sachin is 4 years old, Rehan is 7 years old and Samrina is 9 years old. How much do they each receive?

Key ideas

- Simplify ratios whenever you can.
- To mix in a given ratio both quantities must be multiplied by the same amount.
- To share in a given ratio first add the parts of the ratio together then use the same multiplier for the parts as the total.

7 *Equations with brackets*

You should already know

- how to calculate with negative numbers
- how to solve simple equations.

ACTIVITY 1

You should be able to solve simple equations like these.

a) $5x + 2 = 12$ **b)** $3x - 2 = 13$.

a) $5x + 2 = 12$
$[5x + 2 - 2 = 12 - 2]$ subtracting 2 from each side
$5x = 10$
$[5x \div 5 = 10 \div 5]$ dividing each side by 5
$x = 2$.

b) $3x - 2 = 13$
$[3x - 2 + 2 = 13 + 2]$ adding 2 to each side
$3x = 15$
$[3x \div 3 = 15 \div 3]$ dividing both sides by 3
$x = 5$.

Now try to solve these.

1 $3x + 4 = 16$

2 $5x - 4 = 16$

3 $7x - 3 = 18$

4 $9a + 4 = 40$

5 $8y + 3 = 27$

Equations with brackets

Sometimes the equation formed to solve a problem will involve brackets.

EXAMPLE 1

Solve the equation $2(x + 3) - 6$.
Here are two methods.
Either:

$[2 \times x + 2 \times 3 = 6]$	Expand the brackets.
$2x + 6 = 6$	
$[2x + 6 - 6 = 6 - 6]$	Subtract 6 from each side.
$2x = 0$	
$[2x \div 2 = 0 \div 2]$	Divide each side by 2.
$x = 0$	

Or:

$[2(x + 3) \div 2 = 6 \div 2]$	Divide each side by 2.
$x + 3 = 3$	
$[x + 3 - 3 = 3 - 3]$	Subtract 3 from each side.
$x = 0$	

Exam tip
You may find it easier to do it the first way. By expanding the brackets, you will have an equation like those in revision examples.

Exam tip
This method is usually shorter.

EXERCISE 7.1A

Solve these equations.

1. $2(x + 1) = 10$
2. $3(x + 2) = 9$
3. $4(x - 1) = 12$
4. $5(x + 6) = 20$
5. $2(x - 3) = 7$
6. $3(2x - 1) = 15$
7. $2(2x + 3) = 18$
8. $5(x - 1) = 12$
9. $3(4x - 7) = 24$
10. $2(5 + 2x) = 17$

EXERCISE 7.1B

Solve these equations.

1. $2(x + 4) = 8$
2. $2(x - 4) = 8$
3. $5(x + 1) = 35$
4. $3(x + 7) = 9$
5. $2(x - 7) = 3$
6. $4(3x - 1) = 20$
7. $7(x + 4) = 21$
8. $3(5x - 13) = 21$
9. $2(4x + 7) = 12$
10. $2(2x - 5) = 11$

Equations with the unknown on both sides

Sometimes the equation formed to solve a problem will have the unknown on both sides.

EXAMPLE 2

Solve $2x + 1 = x + 5$.

The first step is the same as before.

$[2x + 1 - 1 = x + 5 - 1]$ — Subtract 1 from each side.

$2x = x + 4.$

Now use the same idea for the x-term on the right-hand side.

$[2x - x = x - x + 4]$ — Subtract x from each side.

$x = 4.$

EXAMPLE 3

Solve $2(3x - 1) = 3(x - 2)$.

Expand the brackets.

$[2 \times 3x + 2 \times (^-1) = 3 \times x + 3 \times (^-2)]$

$6x - 2 = 3x - 6.$

Now use the previous method.

$[6x - 2 + 2 = 3x - 6 + 2]$ — Add 2 to each side.

$6x = 3x - 4.$

$[6x - 3x = 3x - 3x - 4]$ — Subtract $3x$ from each side.

$3x = -4.$

$[3x \div 3 = ^-4 \div 3]$ — Divide each side by 3.

$x = ^-1\frac{1}{3}.$

EXERCISE 7.2A

Solve these equations.

1 $2x - 1 = x + 3$
2 $3x + 4 = x + 10$
3 $5x - 6 = 3x$
4 $4x + 1 = x - 8$

5 $2(x + 3) = x + 7$
6 $5(2x - 1) = 3x + 9$
7 $2(5x + 3) = 5x - 1$

8 $3(x - 1) = 2(x + 1)$
9 $3(3x + 2) = 2(2x + 3)$
10 $3(4x - 3) = 10x - 1$

EXERCISE 7.2B

Solve these equations.

1 $2x + 3 = x + 6$
2 $4x - 1 = 3x + 7$
3 $4x - 3 = x$
4 $5x + 7 = 2x + 16$

5 $2(x - 1) = x + 2$
6 $2(2x + 3) = 3x - 7$
7 $5(3x + 2) = 10x$

8 $3(4x - 1) = 5(2x + 3)$
9 $3(3x + 1) = 5(x - 7)$
10 $7(x - 2) = 3(2x - 7)$

Algebra involves working with letters which stand for unknown numbers. You can always put numbers in when you are not sure.

The two sides of an equation must always be kept equal. Operations carried out to simplify or solve an equation must always be the same for each side.

Everything outside a bracket must be multiplied by everything inside the bracket.

Powers and indices

You should already know

- how to multiply with negative numbers
- how to collect terms.

Indices

Indices or powers are a form of mathematical shorthand:

$3 \times 3 \times 3 \times 3$ is written as 3^4

$2 \times 2 \times 2 \times 2 \times 2 \times 2 \times 2 \times 2$ is written as 2^8

$x \times x \times x \times x \times x$ is written as x^5

EXAMPLE 1

Write these in index form.

a) $5 \times 5 \times 5$ **b)** $a \times a \times a \times a$ **c)** $5 \times 5 \times 6 \times 6 \times 6.$

a) $5 \times 5 \times 5 = 5^3$ **b)** $a \times a \times a \times a = a^4$ **c)** $5 \times 5 \times 6 \times 6 \times 6 = 5^2 \times 6^3.$

EXERCISE 8.1A

Write these in a simpler form, using indices.

1
   a) $2 \times 2 \times 2 \times 2 \times 2$
   b) $6 \times 6 \times 6$
   c) $7 \times 7 \times 7 \times 7 \times 7$
   d) $3 \times 3 \times 3 \times 6 \times 6$
   e) $2 \times 2 \times 3 \times 3 \times 3 \times 3 \times 3$
     $\times 4 \times 4 \times 4.$

2 $b \times b \times b \times b$

3 $p \times p + q \times q$

4 $3a \times 3b$

5 $3a \times 4b$

6 $4a \times b + 2a \times 3b$

EXERCISE 8.1B

Write these in a simpler form, using indices.

1
  a) $5 \times 5 \times 5 \times 5 \times 5$
  b) $9 \times 9 \times 9$
  c) $3 \times 3 \times 3 \times 3 \times 3$
  d) $4 \times 5 \times 5 \times 5 \times 4$
  e) $7 \times 7 \times 8 \times 8 \times 8 \times 9 \times 9$
    $\times 9 \times 9$.

2 $a \times a + b \times b + c \times c$

3 $4 \times 4 \times 4 + 5 \times 5 \times 5$

4 $6 \times 6 - 5 \times 5$

Substituting numbers into a formula

Numbers that can be substituted into a formula can be positive, negative, decimals or fractions.

EXAMPLE 2

a) Find the value of $4x + 3$ when $x = 2$

b) Find the value of $3x^2 + 4$ when $x = 3$

c) Find the value of $2x^2 + 6$ when $x = ^-2$.

a) $4x + 3 = 4 \times 2 + 3$
    $= 17$

b) $3x^2 + 4 = 3(3)^2 + 4$
    $= 3 \times 9 + 4$
    $= 31$

c) $2x^2 + 6 = 2(^-2)^2 + 3$
    $= 2 \times 4 + 3$
    $= 8 + 3$
    $= 11$.

Exam tip

Take special care when negative numbers are involved.

Exam tip

Remember $5b^2$ means $5 \times b \times b$ not $5 \times b \times 5b$.

Exam tip

Work out each term separately and then collect together.

EXERCISE 8.2A

1 Calculate the value of x in each of these expressions by substituting the values of y given.

   a) $x = 5y + 10$ **(i)** $y = 2$
                                     (ii) $y = {}^-3$

   b) $x = 3y^2$ **(i)** $y = 4$
                                       (ii) $y = {}^-2$

   c) $x = 2y^3 + 6$ **(i)** $y = 3$
                                       (ii) $y = {}^-3$

   d) $x = 4y^2 + 6$ **(i)** $y = 0{\cdot}5$
                                       (ii) $y = {}^-0{\cdot}5$.

2 The area of a triangle is given by the formula $A = \frac{1}{2} \times b \times h$.

   Find the area A when

   i) $b = 5$ and $h = 4$

   ii) $b = 4{\cdot}6$ and $h = 5{\cdot}2$.

3 When a stone is dropped from the top of a well the formula $d = 5t^2$ gives the depth of the well if you measure the time, t.

   Find the depth of a well if

   i) $t = 3$ seconds

   ii) $t = 4{\cdot}5$ seconds.

EXERCISE 8.2B

1 Calculate the value of y in each of these formulae by substituting the values of x given.

   a) $y = 3x + 4$ **(i)** $x = 2$
                                         (ii) $x = {}^-2$

   b) $y = x^3 + x^2 + x$ **(i)** $x = 3$
                                         (ii) $x = {}^-2$

   c) $y = 4{\cdot}6x^2 - 4{\cdot}6$ **(i)** $x = 2{\cdot}4$
                                         (ii) $x = {}^-1{\cdot}9$

   d) $y = 10 - 10x^2$ **(i)** $x = 10$
                                         (ii) $x = {}^-2{\cdot}5$

2 A rule for cooking some meats at a temperature of 180°C is $t = 50W + 20$ where W is the weight in kg and t is the time in minutes. Find the time taken to cook a $1{\cdot}2$ kg joint of meat.

3 Find the value of y is $y = x^2 + 2x$ when $x = {}^-1$.

Key ideas

- Remember that $x \times x = x^2$, $x \times x \times x = x^3$.
- When squaring a negative number the answer is positive e.g. $({}^-2)^2 = {}^-2 \times {}^-2 = 4$.

Chapter 8 *Powers and indices*

9 Circles

- the terms area, perimeter, circumference, radius, diameter, chord
- the units for area: cm², m².

You need to be able to identify the parts of a circle

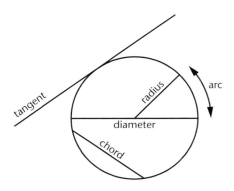

Circumference is the distance all the way round – the perimeter of the circle.

ACTIVITY 1

Using a tape measure, or a piece of string and a ruler, measure the circumference and diameter of eight circular objects. Make sure you use the same units for both measurements.

Copy this table and complete it. For the fourth column use your calculator to work out the value of: circumference $C \div$ diameter d

Name of object	Circumference (C)	Diameter (d)	C ÷ d

The numbers in the fourth column of the table should all be about 3.

The approximate relationship $C \approx 3 \times d$ has been known for thousands of years. In fact, very accurate calculations have shown that instead of 3 the multiplier should be approximately 3·142. Even more accurate calculations have found this number to hundreds of decimal places. Because the number is a never-ending decimal, it is denoted by the Greek letter π.

> **Exam tip**
>
> Can you think of a reason why $C \div d$ should be constant? It is because all circles are similar (the same shape), so the ratio of corresponding lengths will be constant. Try it with the sides of a rectangle 2 units × 1 unit, and similar rectangles with sides of 4 units × 2 units, 6 units × 3 units, and so on.

Your calculator has the number which π represents stored in its memory.

Find the π button on your calculator and write down all the digits it shows. On some calculators you may have to press = after pressing the π button.

The relationship between circumference and diameter can now be written as:

$C = \pi \times d$

Remember, when two letters are written next to each other in algebra, it means multiply. The formula can be written shortly as $c = \pi d$.

If you know the radius of the circle instead of the diameter, use the fact that the diameter is double the radius.

$d = 2r$

The formula then becomes:

$C = \pi \times 2r$

Since multiplication can be done in any order, this can be written as $C = 2 \times \pi \times r$ or, using the shorter version:

$C = 2\pi r$

> **Exam tip**
>
> If you are using your calculator always use the π button, rather than another approximation. If you are sitting the non-calculator paper you may be asked to leave answers in terms of π. In this case, in Example 1, the answer could be given as 5π.
> If not, the approximation 3·14 or 3·142 will be sufficient, you can always do a rough check of your calculations using $\pi = 3$.

EXAMPLE 1

A circle has a diameter of 5 cm. Find its circumference.

Circumference = πd

$= \pi \times 5$

$= 15.707\,96\ldots$

$= 15.7$ cm (to 1 d.p.)

EXAMPLE 2

A circle has a radius of 8 cm. Find its circumference.

Circumference = $2\pi r$

$= 2 \times \pi \times 8$

$= 50.265\ldots$

$= 50.3$ cm (to 1 d.p.)

EXAMPLE 3

A circle has a circumference of 20 m. Find its diameter.

Circumference = πd so

$20 = \pi \times d$.

Solving the equation gives: $d = 20 \div \pi$

$= 6.366\ldots$

$= 6.37$ m (to 2 d.p.)

EXERCISE 9.1A

1 Calculate the circumference of the circles with these diameters, giving your answers correct to one decimal place.

   a) 12 cm **b)** 9 cm **c)** 20 m
   d) 16·3 cm **e)** 15·2 m.

2 Find the circumference of circles with these radii, giving your answers correct to one decimal place.

   a) 5 cm **b)** 7 cm **c)** 16 m
   d) 18·1 m **e)** 5·3 m.

3 The centre circle on a football pitch has a radius of 9·15 metres. Calculate the circumference of the circle.

4 The diagram shows a wastepaper bin in the shape of a cylinder. Anthea is going to decorate it with braid around the top rim. Calculate the length of braid she needs.

30 cm

5 The circumference of a circle is 50 cm. Calculate the diameter of the circle.

EXERCISE 9.1B

1 Calculate the circumferences of circles with these diameters, giving your answers correct to one decimal place.

   a) 25 m   **b)** 0·3 cm   **c)** 17 m
   d) 5·07 m   **e)** 6·5 cm

2 Find the circumference of circles with these radii, giving your answers correct to one decimal place.

   a) 28 cm   **b)** 3·2 cm   **c)** 60 m
   d) 1·9 m   **e)** 73 cm

3 The radius of the earth at the equator is 6378 km. Calculate the circumference of the earth at the equator.

4 Bijan and Claire have a ride on the roundabout at a fair.
Bijan rides on a motorcycle which is 2·5 metres from the centre of the roundabout. Claire rides on a horse which is 3 metres from the centre.

On each ride the roundabout goes round ten times.
How much further does Claire travel than Bijan?

5 A circular racetrack is 300 metres in circumference. Calculate the diameter of the racetrack.

Area of a circle

The diagram shows a circle, radius r, divided into 36 sectors of centre angle 10°.

The next diagram shows these sectors cut out and rearranged.

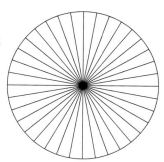

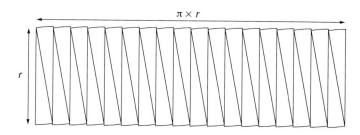

This shape is now very close to a rectangle.

The length of this rectangle will be approximately half the circumference. That is $2\pi r \div 2 = \pi r$.

The width of the rectangle is r.

The area of this rectangle = length $\times$ width = $\pi r \times r = \pi r^2$.

If you take more sectors, with smaller angles, the shape is even closer to a rectangle.

The formula for the area of a circle is therefore:

area of circle $A = \pi r^2$

Exam tip

One of the most common error is to mix up diameter and radius. Every time you do a calculation make sure you have used the right one.

Exam tip

Make sure you can use the π button and the square button on your calculator so you do not have to write down long decimals before the final answer.

EXAMPLE 4

The radius of a circle is 6 cm. Calculate the area of the circle.

$A = \pi r^2 = \pi \times 6^2 = \pi \times 36 = 113 \cdot 1 \, \text{cm}^2$ (to 1 d.p.).

EXAMPLE 5

The radius of a circle is 4·3 m. Calculate the area of the circle.

$A = \pi r^2 = \pi \times 4.3^2 = 58 \cdot 1 \, \text{m}^2$ (to 1 d.p.).

EXAMPLE 6

The diameter of a circle is 18·4 cm. Calculate the area of the circle.

$r = 18 \cdot 4 \div 2 = 9 \cdot 2 \, \text{cm}$.

$A = \pi r^2 = \pi \times 9 \cdot 2^2 = 266 \, \text{m}^2$ (to the nearest m^2).

EXAMPLE 7

The radius of a circle is 5 m. Calculate the area of the circle, leaving your answer as a multiple of π.

$A = \pi r^2 = \pi \times 5^2 = \pi \times 25 = 25\pi \, \text{m}^2$.

EXERCISE 9.2A

In all of these questions, make sure you state the units of your answer.

1 Find the areas of circles with these radii.

   a) 4 cm   **b)** 16 m   **c)** 11·3 m   **d)** 13·6 m   **e)** 8·9 cm.

2 The radius of a circular fish pond is 1·5 m. Find the area of the surface of the water.

3 The diameter of a circular table is 0·8 m. Find the area of the table.

4 To make a table mat, a circle of radius 12 cm is cut from a square of side 24 cm as shown in the diagram on the right. Calculate the area of material that is wasted.

5 The radius of the circular face of a church clock is 1·2 m. Calculate the area of the clock face.

6 Use your calculator to find the area of a circle with radius 6·8 cm. Without using your calculator do an approximate calculation to check your answer.

Exam tip

In the last two sections all the questions in the exercise have been about circumference, or all of them have been about area, so there has been no problem about which formula to use. In an examination you will have to choose the formula. If you have trouble remembering which of the circle formulae gives circumference and which gives area, remember r^2 must give the units cm^2 or m^2 and so πr^2 must be the area formula.

EXERCISE 9.2B

In all of these questions, make sure you state the units of your answer.

1 Find the areas of these circles.

a) 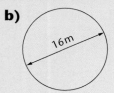 3 cm

b) 16 m

c) 5·3 cm

d) 26·4 cm

e) 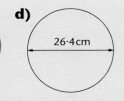 2·3 m

2 A circular mouse mat has a radius of 9 cm. Find the area of the mouse mat.

3 According to the *Guiness Book of Records*, the largest pizza ever made was 37·4 m in diameter. What was the area of the pizza?

4 A square has a side of 3·5 cm and a circle has a radius of 2 cm. Which has the bigger area? You must show your calculations.

5 Charlie is making a circular lawn with a radius of 15 m. The packets of grass seed say 'sufficient to cover 50 m^2'. How many packets will she need?

6 The diagram shows a child's plastic ring. The diameter of the large circle is 30 cm. The diameter of the small circle is 20 cm. Calculate the area of the ring (shaded).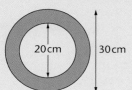

20 cm   30 cm

Key ideas

● The circumference of a circle is given by **2πr** or **πd**.
● The area of a circle is given by **πr²**.
● Use the **π** button on a calculator if you are able to; if not use the approximation **3·14** or **3·142**.

Revision exercise

1 Change these fractions to decimals, giving your answer to three decimal places where appropriate.

   a) $\frac{5}{16}$ **b)** $\frac{4}{7}$ **c)** $\frac{7}{40}$ **d)** $\frac{4}{15}$.

2 Change the decimals you found in question 1 into percentages. Give your answers to one decimal place.

3 Find £1·50 as a percentage of £21. Give your answer to the nearest 1%.

4 Find 80 cm as a percentage of 6 m. Give your answer to one decimal place.

5 Karl cuts 20 cm from a piece of wood 1·6 m long. By what percentage has he shortened the piece of wood?

6 The audience for a TV soap increased from 8 million to 10 million. What percentage increase is this?

7 The number of pupils in a school went down from 850 to 799. What percentage reduction is this?

8 In a sale all prices are reduced by 15%. Find the new price of a pair of trainers that originally cost £65.

9 Each year the value of an antique increases by 20% of its value at the beginning of the year. If it was worth £450 on 1 January 1996, what was it worth on 1 January 1997? What was it worth on 1 January 1999?

10 Write each ratio in its simplest form.

   a) $9:2$ **b)** $25:5$ **c)** $60:40$
   d) $39:13$ **e)** $60:144$

11 Write each ratio in its simplest form:

   a) $70\,\text{cl}:1$ litre   **b)** £2 : 80p
   c) 9 hours : $1\frac{1}{2}$ hours
   d) $9\,\text{kg}:150\,\text{g}$

12 A shade of orange paint is made by mixing red and yellow paint in the ratio 1 : 3.

   a) How many tins of yellow paint would you mix with four tins of red?

   b) How many tins of red paint would you mix with eighteen tins of yellow?

13 The instructions for making flaky pastry say, 'First mix flour and fat in the ratio 8 parts to 5 parts.' How much fat should I mix with 480g of flour?

14 Laura and Marie share a flat. They agree to share the rent in the same ratio as their wages. Laura earns £600 per month and Marie earns £800 per month. If the rent is £420 per month, how much do they each pay?

15 Solve these equations.
  a) $2(2x - 1) = 3(x - 2)$
  b) $3(5x + 2) = 7(2x + 1)$
  c) $4x + 7 = 3(4 + x)$
  d) $5(2x - 3) = 15$
  e) $3(3x - 2) = 2(2x - 3)$.

16 Solve these equations.
  a) $\frac{1}{2}x + 1 = 4$
  b) $3 - x = 2x - 3$
  c) $2(3x - 2) = 4(2x + 3) - x$
  d) $3(7 - x) = 4(1 - 2x)$
  e) $\frac{2}{3}x = \frac{3}{2}$

17 Simplify:
  a) $3 \times 3 \times 3 \times 3 \times 3 \times 3 \times 3$
  b) $6 \times 6 \times 7 \times 7 \times 6$
  c) $a \times a \times a \times a \times a$.

18 If $x = {}^{-}3$ find the value of
  a) $2x^2$ **b)** $3x^3$.

19 If $a = {}^{-}2$ and $b = 3$ find the value of $3b^2 - 2a^3$.

20 Find the circumference of circles with:
  a) diameter 6 cm
  b) radius 10 cm
  c) diameter 4 cm

21 Find the area of circles with:
  a) diameter 4 cm
  b) radius 3 cm
  c) radius 12 cm.

22 The distance round a circular flower bed is 36 metres.
  a) Find the radius of the flower bed.
  b) Find the area of the flower bed.

10 Scatter diagrams

You should already know

- **know how to plot points**
- **understand the mean.**

Scatter diagrams

Scatter diagrams (also called scatter graphs) are used to investigate any possible link or relationship between two features or variables. Values of the two features are plotted as points on a graph. If these points tend to lie in a straight line then there is a relationship or correlation between the two features.

EXAMPLE 1

The table below shows the mean annual temperature for 12 cities which lie north of the equator.

City	Latitude (degrees)	Mean temperature (°C)
Bombay	19	31
Casablanca	34	22
Dublin	53	13
Hong Kong	22	25
Istanbul	41	18
St Petersburg	60	8
Manila	15	32
Oslo	60	10
Paris	49	15
London	51	12
New Orleans	30	22
Calcutta	22	26

When these data are plotted, as shown on the next page, there appears to be a relationship between temperature and latitude: the farther north the city the lower the temperature will tend to be.

Example 1 cont'd

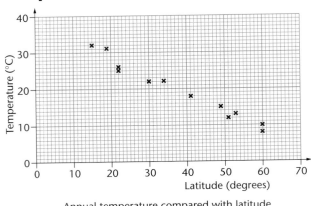

Annual temperature compared with latitude

EXAMPLE 2

This table shows the percentage marks gained in a maths test and an English test by the same 14 students.

Maths	62	53	23	61	25	46	48	49	60	61	61	69	85	48
English	52	53	45	57	48	49	53	53	56	58	59	54	62	53

The scatter graph showing these data suggests that there is a relationship between the results of the two subjects: the higher the maths mark, the higher the English mark.

A fifteenth student took the maths test but missed the English test. Her maths score was 40%. The scatter graph can be used to estimate her English score as around 50%. Check that you agree.

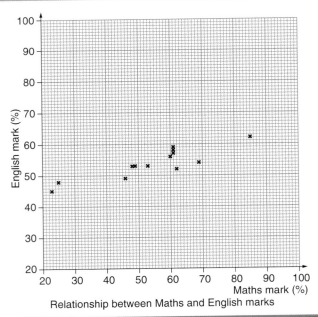

Relationship between Maths and English marks

1 The marks of ten students in the two papers of a French exam are shown in the table.

Paper 1	20	32	40	45	60	67	71	80	85	91
Paper 2	15	25	40	40	50	60	64	75	76	84

Plot these marks on a scatter diagram.

A student scored 53 marks on paper 1. What would you guess his likely mark to be on paper 2?

2 In the Key Stage 2 maths National Tests the pupils take two written tests, test A and test B, and a mental mathematics test. The marks for the 26 pupils in class 6 are shown in the table.

Test A	Test B	Mental mathematics	Test A	Test B	Mental mathematics
28	32	16	9	6	5
36	37	19	18	33	12
6	5	3	35	36	18
17	18	5	12	17	6
24	31	16	26	33	15
21	29	17	9	11	4
24	27	14	27	31	19
12	11	9	13	22	9
19	21	17	35	37	18
27	37	16	15	23	12
30	28	16	8	12	3
13	22	6	25	37	19
19	29	17	14	24	19

Draw scatter diagrams to compare the marks between test A and the mental test score, and between test A and test B.

Exercise 10.1A cont'd

3 The table below gives the height and the resting pulse rate for 18 people.

Draw a scatter diagram and investigate if there is a relationship between height and pulse rate.

Can you estimate your graph the pulse of someone 185 cm tall? or someone 140 cm tall?

Height (cm)	Pulse rate (per minute)	Height (cm)	Pulse rate (per minute)
160	68	165	84
162	64	172	116
180	80	163	95
173	92	168	90
170	80	182	76
163	80	170	84
148	82	155	80
160	84	175	104
180	90	180	68

EXERCISE 10.1B

1 Twelve people were chosen at random and asked to solve the problem. The time each person took and their age were recorded.

Age	14	45	18	74	60	23	21	56	20	39	30	40
Time	20	12	12	23	15	6	5	18	6	11	7	10

a) Draw a scatter diagram
b) comment on the graph

2 Sanjay is convinced that the more chocolate bars he eats the heavier he will become.

Sketch a scatter graph to show this.

Correlation

The table below shows the amount of ice-cream sold by an ice-cream seller in ten days last summer.

Number of hours of sunshine	3	6	11	2	0	7	2	12	7	5
Number of ice-creams sold	120	200	360	100	50	250	150	470	330	230

The graph below shows this information plotted on a scatter diagram or scatter graph.

The number of hours of sunshine is plotted as the x-coordinate and the number of ice-creams sold is plotted as the y-coordinate.

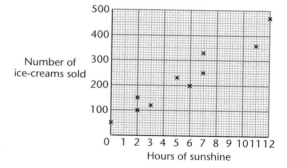

From the graph, it can be seen that, in general, the more hours of sunshine there were, the more ice-creams were sold.

This is an example of positive correlation.

Although the points are not exactly in a straight line, nevertheless there is a trend that the further to the right on the graph the higher the point is.

In graphs such as these the nearer the graph is to a straight line, the better the correlation is.

Examples of graphs showing positive correlation.

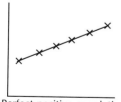
Perfect positive correlation

Strong positive correlation

Weak positive correlation

Chapter 10 *Scatter diagrams*

If there is no correlation the scatter diagram looks like this.

A shopkeeper in the same town as the ice-cream seller noted how many umbrellas were sold in the same ten days. This is the table.

Number of hours of sunshine	3	6	11	2	0	7	2	12	7	5
Number of umbrellas sold	6	5	2	9	11	4	8	0	5	7

The scatter diagram for this information looks like this.

The number of hours of sunshine is plotted as the x-coordinate and the number of umbrellas sold is plotted as the y-coordinates.

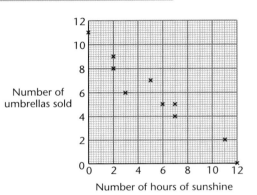

Here the trend is the other way round. In general, although the points are not exactly in a straight line, the more hours of sunshine there are the fewer umbrellas are sold.

This is an example of negative correlation.

Examples of graphs showing negative correlation.

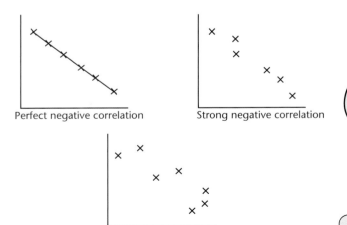

Perfect negative correlation   Strong negative correlation

Weak negative correlation

> **Exam tip**
>
> When commenting on a scatter diagram it is better (and quicker) to use the correct terms such as 'strong positive correlation' rather than using phrases like 'the more hours of sunshine, the more ice-creams are sold'.

Lines of best fit

Look again at the graph for ice-cream and hours of sunshine.

A straight line has been drawn on it, passing through the cluster of points. There are as many points above the line as there are below it. This is the 'best' straight line that can be drawn to show the trend of the points. It is called the line of best fit.

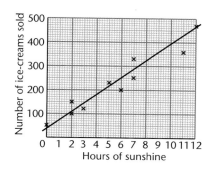

> The line of best fit should reflect the slope of the points and have approximately the same number of points on either side.

It should ignore any points that obviously do not fit the trend. These are called outliers.

A line of best fit should not be attempted if there is little or no correlation.

The line of best fit can be used to estimate values that are not in the original table. For example, you could estimate that for 5 hours of sunshine 210 ice-creams would be sold.

If the line of best fit is used to estimate values it must be recognised that:

● if the correlation is not good the estimate will probably not be a very good one

● estimates should not be made too far beyond the range of the given points. For example, in the above case, estimates should not be made for 15 hours of sunshine.

Examples of bad 'lines of best fit

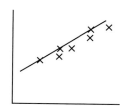

Fault: Slope about right but does not have the same number of points on either side.

Fault: Same number of points either side but slope wrong.

Exam tip

When drawing a line of best fit:
1 put your ruler on the graph at the right slope
2 slide the ruler, keeping it at the same slope, until you have approximately the same number of points on either side.
Remember, lines of best fit do not necessarily go through the origin.
It may be easier to hold your ruler on its edge, vertical to the page so you can see both sides of the line.

EXERCISE 10.2A

1 The table on the right shows the temperature at noon, (local time), one day in Autumn for 15 cities together with their latitudes.

City	Latitude	Temperature (°C)
Algiers	37	23
Berlin	53	13
Boston	43	19
Cairo	30	31
Dublin	54	11
Helsinki	60	6
Iverness	58	10
London	52	12
Mexico City	19	26
New York	41	19
Paris	49	13
Tangiers	36	20
Tel Aviv	32	28
Tokyo	36	24
Washington	39	22

  a) Draw a scatter diagram to show this information, (use latitude on the horizontal axis)

  b) Comment on the graph

2 Eight people were chosen at random and aked to state their age, weight and height. The table shows the data.

Person	A	B	C	D	E	F	G	H
Age, (years)	28	32	42	40	53	50	36	70
Height, (m)	2·0	1·6	1·7	1·5	1·9	1·9	1·9	1·8
Weight (kg)	74	59	59	50	62	68	71	56

  a) draw a scatter diagram to show age and weight
  b) draw a scatter diagram to show height and weight
  c) comment on anything you notice from the two diagrams

3 The table below shows the scores in 10 football matches. Show the information on a scatter diagram and comment on any pattern you see.

Match	1	2	3	4	5	6	7	8	9	10
Goals scored by home side	3	4	0	0	1	2	2	3	2	1
Goals scored by away side	0	1	2	3	2	2	0	0	5	3

EXERCISE 10.2B

1 The table shows the heights and ages for a group of pupils.

Age, years	10	10·5	12	12·25	12·5	12·75	13·5
height, cm	115	134	153	134	150	160	140

Age, years	13·25	14·25	14·5	12·25
height, cm	166	188	174	184

a) Show this data on a scatter graph.
b) Use your graph to estimate the age of a pupil who is 144 cm tall.

2 Tony asked his friends how many packets of crisps they ate in a week and how many cans of fizzy drink they drank in the same week. His results are as follows:

cans of drink	6	8	1	0	9	6	3	3	5	4
packets of crisps	5	6	1	2	7	4	2	3	4	4

a) Show this on a scatter graph.
b) Describe the relationship between the number of cans of drink and the number of packets of crisps eaten.
c) Estimate the number of packets of crisps eaten during a week by a student who drank 7 cans of coke.

3 The table below shows the mathematics and English marks of ten students

Maths	93	75	40	64	80	63	94	33	40	55
English	64	54	23	70	32	50	80	25	32	45

a) Draw a scatter graph to show this information, with the maths mark along the horizontal axis.
b) Draw a line of best fit.
   Use your line to estimate
   i) the English mark of a student who gained 70 marks in mathematics
   ii) the maths mark of a student who gained 40 marks in English.

Key ideas

A scatter graph shows if there is a relationship between two sets of data.

11 Multiplication and division of fractions

Multiplying fractions

You think of a fraction as the number 1 multiplied by the numerator and divided by the denominator. For example, $\frac{2}{3} = 1 \times 2 \div 3$ or $\frac{1}{1} \times \frac{2}{3} = \frac{1 \times 2}{1 \times 3}$.

To multiply fractions, multiply the numerators and multiply the denominators, then simplify if possible. If the fractions are mixed numbers change them to top-heavy fractions and then multiply.

> **Exam tip**
>
> Multiplying a fraction by a whole number is easy. One way is to think of the whole number as a fraction with a denominator of 1 and carry on as before.

EXAMPLE 1

Work these out.

a) $\frac{3}{5} \times \frac{1}{2}$ **b)** $\frac{3}{8} \times 2$ **c)** $1\frac{4}{7} \times 3$

a) $\frac{3}{5} \times \frac{1}{2} = \frac{3}{10}$

Multiply the numerators and multiply the denominators. The answer is already in its lowest terms.

b) $\frac{3}{8} \times 2 = \frac{3}{8} \times \frac{2}{1} = \frac{6}{8} = \frac{3}{4}$

c) $1\frac{4}{7} \times 3 = \frac{11}{7} \times \frac{3}{1} = \frac{33}{7} = 4\frac{5}{7}$

> **Exam tip**
>
> Note: $\frac{1}{2} \times \frac{1}{3} = \frac{1}{6}$
> A common error is to multiply 1×1 and get 2.

EXERCISE 11.1A

Work these out.

1 $\frac{1}{2} \times 4$ **4** $\frac{3}{4} \times 12$ **7** $\frac{2}{3} \times 5$

2 $7 \times \frac{1}{2}$ **5** $\frac{3}{4} \times \frac{1}{2}$ **8** $24 \times \frac{5}{12}$

3 $9 \times \frac{1}{3}$ **6** $\frac{1}{3} \times \frac{3}{4}$

Remember that, when simplifying or cancelling, you can just cross out the numbers you are cancelling and write in the quotients, for example, $\frac{3}{8} \times 2 = {}_4\frac{3}{\cancel{8}} \times \frac{\cancel{2}^1}{1} = \frac{3}{4}$.

> ### Exam tip
> When cancelling, cancel a term in the numerator with one in the denominator.

Dividing fractions

Dividing a fraction by a whole number is straightforward – as with multiplication think of the whole number as a fraction with a denominator of 1 and carry on as before.

Multiplying by $\frac{1}{2}$ is the same as dividing by 2, so dividing by $\frac{1}{2}$ is the same as multiplying by 2.

This can be extended, for example, $4 \div \frac{2}{3} = 4 \times \frac{3}{2} = \frac{12}{2} = 6$.

When dividing fractions, turn the second fraction upside-down and then multiply.

EXAMPLE 2

Work these out.

a) $\frac{8}{9} \div \frac{1}{3}$ **b)** $\frac{8}{9} \div 2$ **c)** $\frac{1}{2} \div \frac{1}{3}$

a) $\frac{8}{9} \div \frac{1}{3} = \frac{8}{9} \times 3$

          $= \frac{8}{3} \times 1$ Turn the second fraction upside down and multiply

          $= \frac{8}{3}$ Cancel 9 and 3 by 3 and change to a mixed number

          $= 2\frac{2}{3}$

b) $\frac{8}{9} \div 2 = \frac{8}{9} \times \frac{1}{2}$

          $= \frac{4}{9} \times 1$ Cancel 8 and 2 by 2

          $= \frac{4}{9}$

c) $\frac{1}{2} \div \frac{1}{3} = \frac{1}{2} \times \frac{3}{1}$

          $= \frac{3}{2}$

          $= 1\frac{1}{2}$

Chapter 11 *Multiplication and division of fractions*

EXERCISE 11.1B

Work these out.

1 $\frac{1}{2} \div 4$ **4** $12 \div \frac{3}{4}$ **7** $\frac{2}{3} \div 5$

2 $7 \div \frac{1}{2}$ **5** $\frac{3}{4} \div \frac{1}{2}$ **8** $\frac{1}{3} \div \frac{3}{4}$

3 $9 \div \frac{1}{3}$ **6** $\frac{2}{3} \div 3$

EXERCISE 11.2A

Work these out.

1 $\frac{2}{3} \times \frac{1}{4}$ **6** $\frac{4}{9} \div 2$

2 $\frac{2}{3} \times 4$ **7** $\frac{3}{8} \div \frac{1}{4}$

3 $\frac{4}{9} \times \frac{1}{2}$ **8** $\frac{4}{5} \div 4$

4 $\frac{2}{3} \times \frac{1}{3}$ **9** $\frac{2}{3} \div \frac{1}{3}$

5 $\frac{4}{9} \times 2$ **10** $1\frac{2}{5} \times 4$

EXERCISE 11.2B

Work these out.

1 $\frac{1}{2} \times \frac{5}{6}$ **6** $\frac{2}{3} \div \frac{1}{6}$

2 $\frac{4}{5} \times 3$ **7** $\frac{5}{6} \div 10$

3 $\frac{3}{4} \times \frac{1}{8}$ **8** $\frac{7}{9} \div \frac{1}{9}$

4 $\frac{7}{9} \times \frac{1}{7}$ **9** $\frac{3}{4} \div \frac{1}{8}$

5 $\frac{3}{5} \times \frac{1}{12}$ **10** $1\frac{2}{5} \div 4$

Key ideas

- When cancelling or simplifying a fraction be sure to cancel a term in the numerator with one in the denominator.
- Remember to multiply numerators together and multiply denominators together.
- Division of fractions means turn the second fraction upside down and multiply.

12 Negative numbers and decimals

Adding and subtracting negative numbers

Adding a negative number is the same as subtracting a positive number.

$4 + (^-2) = 4 - 2 = 2$.

Subtracting a negative number is the same as adding a positive number.

$4 - (^-2) = 4 + 2 = 6$.

When the numbers to be added are both negative, add them and put the negative sign in front.

$^-4 - 5 = ^-9$.

If they have different signs, subtract them and give the answer the same sign as the larger number.

$^-4 + 6 = 2$

$^-5 + 2 = ^-3$.

When adding or subtracting negative and positive numbers, it is best to:

total the positive numbers

total the negative numbers separately

then find the difference between the two totals,

remembering to give the answer the correct sign.

You can already multiply and divide negative numbers. Now the steps can be combined.

EXAMPLE 1

Work these out.

a) $(^-3 \times ^-4) + (^-2 \times 3)$

b) $^-5 + 4 - 6 + 7 + 5 + 1 - 3 - 5 - 2$

c) $\dfrac{5 \times ^-4 + 3 \times ^-2}{^-6 + 4}$.

a) $(^-3 \times ^-4) + (^-2 \times 3) = (+12) + (^-6) = 12 - 6 = +6 = 6$

b) $^-5 + 4 - 6 + 7 + 5 + 1 - \times - 3 - 2 = +17 - 21 = ^-4$

c) $\dfrac{5 \times ^-4 + 3 \times ^-2}{^-6 + 4} = \dfrac{^-20 + (^-6)}{^-2} = \dfrac{^-26}{^-2} = 13$.

EXERCISE 12.1A

Work these out.

1 $(^-4 \times ^-3) - (^-2 \times +1)$

2 $(^-7 \times ^-2) + (4 \times ^-2)$

3 $(^-15 \div 2) - (4 \times ^-6)$

4 $^-4 + 3 + 2 + 3 + 4 - 5 - 6 - 9 + 1$

5 $\dfrac{^-2 + 12}{^-5}$

6 $\dfrac{^-4 \times ^-3}{^-4 + 3}$

7 $\dfrac{^-4 \times 5}{^-6 + 4}$

8 $(16 \div 2) - (^-2 \times 4)$

EXERCISE 12.1B

Work these out.

1 $(^-2 \times +3) + (^-3 \times +4)$

2 $(^-1 \times ^-4) + (^-7 \times ^-8)$

3 $(24 \div ^-3) - (^-5 \times ^-4)$

4 $^-6 - 2 - 3 + 5 - 7 + 4 - 2 + 8$

5 $\dfrac{^-3 + 7}{^-2}$

6 $\dfrac{^-7 \times ^-12}{^-8 + 4}$

7 $\dfrac{8 \times ^-6}{^-4 + ^-8}$

8 $\dfrac{9 \times 4}{^-3 \times ^-6}$

Decimals – a reminder

You need to remember:

(i) the links between decimals and place value

and (ii) that decimals are really fractions with denominators of 10 or 100 or 1000 etc:

hundreds	tens	ones	.	tenths	hundredths
4	3	5	.	2	7

4 hundreds + 3 tens + 5 ones + 2 tenths + 7 hundredths
$= 400 + 30 + 5 + \frac{2}{10} + \frac{7}{100}$
$= 435.27$

Addition and subtraction of decimals

When adding or subtracting decimals without using a calculator take care to line up the decimal points.

EXAMPLE 2

$16{\cdot}45 + 2{\cdot}62 = $
$$\begin{array}{r} 16{\cdot}45 \\ + 2{\cdot}62 \\ \hline 19{\cdot}07 \end{array}$$

Remember every time you add and subtract amounts of money you are dealing with decimals; remember too that you should write three pounds twenty pence as £3·20 and not £3·2 so, for example, £3·22 + £2·18 = £5·40 not £5·4.

EXERCISE 12.2A

Work out the answers to the following questions:

1 **a)** £2·10 + £3·45
   b) £5·78 + £2·82
   c) £7·15 − £6·13
   d) £34·02 + £14·89
   e) £123·67 − £65·77.

2 Anna buys a T-shirt costing £8·99 and a jumper costing £18·99. How much change does she get from £30?

3 Work out the value of the following.
   a) 1·6 + 3·4   **b)** 4·9 + 5·21
   c) 17·77 + 19·54 **d)** 10·78 − 4·9
   e) 21·99 − 11·9   **f)** 5·53 − 3·09.

4 Class 7 are collecting money each week for a charity. The first 4 weeks totals are:

   £12·56, £14·66, £18·13, £11·82,

   How much more do they need to raise to reach their target of £100?

EXERCISE 12.2B

Work out the answers to the following questions:

1 a) £4·37 + £6·53
   b) £15·78 + £9·89
   c) £6·18 + £6·32
   d) £15·42 – £9·34
   e) £19·21 – £17·04

2 Helen buys a packet of sandwiches costing £1·60, a bar of chocolate costing 45p and a drink costing 65p. How much change will she get from £5?

3 Work out the value of the following:

   a) 4·9 + 5·6   b) 34·94 + 7·62
   c) 24·24 + 16·16   d) 8·6 – 3·4
   e) 9·42 – 8·57   f) 24·2 – 4·63.

4 Kate buys a 4 m length of material. She makes two curtains, each of which is 1·8 m long. How much material does she have left?

Multiplication and division of decimals

Remember when multiplying decimals by 10 then all the digits move one place to the left so 435·27 × 10 becomes 4352·7.

When multiplying by 100 the digits move two places to the left and so on. However we usually talk about the decimal point moving because it is easier to move the decimal point when working on paper.

thousand	hundreds	tens	ones	•	tenths	hundredths
		4	3	5	•	2	7
4	3	5	2		7	

EXAMPLE 3

a) 0·35 × 10   b) 4·567 × 100.

a) 0·35 × 10 = 3·5, the decimal point has moved 1 place to the right – this is the same as the number moving one place to the left.

b) 4·567 × 100 = 456·7, the decimal point has moved 2 places to the right.

EXAMPLE 4

a) 3·5 ÷ 10   b) 456·7 ÷ 100.

a) 3·5 ÷ 10 = 0·35   b) 456·7 ÷ 100 = 4·567.

When dividing move the decimal point to the left.

EXERCISE 12.3A

Work out.

1 **a)** $2 \cdot 4 \times 10$ **b)** $3 \cdot 09 \times 10$
    c) $32 \cdot 6 \times 10$ **d)** $1 \cdot 3 \times 100$
    e) $32 \cdot 45 \times 100$ **f)** $0 \cdot 098 \times 100$.

2 **a)** $45 \cdot 1 \div 10$ **b)** $32 \cdot 78 \div 100$
    c) $45 \cdot 45 \div 100$ **d)** $0 \cdot 8 \div 10$
    e) $0 \cdot 03 \div 100$ **f)** $46 \cdot 9 \div 1000$.

EXERCISE 12.3B

Work out.

1 **a)** $3 \cdot 6 \times 10$ **b)** $5 \cdot 46 \times 10$
    c) $19 \cdot 73 \times 100$ **d)** $14 \cdot 62 \times 10$
    e) $23 \cdot 49 \times 100$ **f)** $2 \cdot 36 \times 100$.

2 **a)** $37 \cdot 9 \div 10$ **b)** $54 \cdot 63 \div 100$
    c) $42 \cdot 42 \div 1000$ **d)** $19 \cdot 29 \div 100$
    e) $37 \cdot 84 \div 10$ **f)** $426 \cdot 4 \div 1000$.

When multiplying two decimal numbers the method we use is:

    ●   first 'ignore' the decimal points,

    ●   then do the multiplication

and   ●   finally count up the number of decimal figures in the question numbers – the total will give the number of decimal figures in the answer number.

EXAMPLE 5

a) $0 \cdot 4 \times 0 \cdot 5$ **b)** $1 \cdot 23 \times 1 \cdot 2$.

a) $0 \cdot 4 \times 0 \cdot 5$ is calculated as $4 \times 5 = 20$.

There are two decimal figures in the question numbers, (4 and 5) so there are two in the answer. Therefore $0 \cdot 4 \times 0 \cdot 5 = 0 \cdot 20$.

b) $1 \cdot 23 \times 1 \cdot 2$

First work out $123 \times 12 = 1476$.

Then put in the decimal point.

There are three decimal figures the '23' in $1 \cdot 23$ and the '2' in $1 \cdot 2$ so there are three decimal figures in the answer.

Thus $1 \cdot 23 \times 1 \cdot 2 = 1 \cdot 476$.

EXERCISE 12.4A

Work out.

1 **a)** $2 \times 4 \cdot 6$ **b)** $4 \times 7 \cdot 9$
    c) $3 \times 4 \cdot 5$ **d)** $12 \times 3 \cdot 2$
    e) $6 \times 7 \cdot 9$ **f)** $5 \times 45 \cdot 2$.

2 **a)** $200 \times 0 \cdot 56$ **b)** $8 \cdot 6 \times 40$
    c) $5 \cdot 9 \times 200$ **d)** $0 \cdot 07 \times 300$
    e) $0 \cdot 01 \times 4000$ **f)** $600 \times 4 \cdot 5$.

3 Given that $63 \times 231 = 14\,553$ write down the answers to
    a) $6 \cdot 3 \times 2 \cdot 31$ **b)** $63 \times 23 \cdot 1$
    c) $0 \cdot 63 \times 23 \cdot 1$ **d)** $63 \times 0 \cdot 231$
    e) $6 \cdot 3 \times 23\,100$.

Chapter 12 *Negative numbers and decimals*

EXERCISE 12.4B

Work out.

1 **a)** 14×6.4 **b)** 9×7.3
    c) 8.4×6 **d)** 13.3×5.
2 **a)** 100×0.49 **b)** 33.3×40
    c) 200×0.6 **d)** 400×3.8.
3 Given that $12.4 \times 8.5 = 105.4$
write down the answers to.

    a) 124×8.5 **b)** 12.4×0.85
    c) 0.124×8.5 **d)** 1.24×8.5
    e) 0.124×850.

When dividing decimals then we work out the answers as follows.

(i) If dividing by a whole number then line up the decimal point in the answer with the decimal point in the question.

EXAMPLE 6

$$75.5 \div 5 = 5)\overline{75.5} \quad \begin{array}{c} 15.1 \end{array}$$

EXERCISE 12.5A

Work out.

1 **a)** $40.5 \div 5$ **b)** $210.3 \div 3$ **c)** $2.644 \div 4$ **d)** $36.33 \div 3$.
2 **a)** $3.2 \div 4$ **b)** $39.6 \div 6$ **c)** $12.5 \div 5$ **d)** $49.63 \div 7$.

EXERCISE 12.5B

Work out:

1 **a)** $81.9 \div 9$ **b)** $64.8 \div 8$
    c) $124.4 \div 4$ **d)** $64.4 \div 7$.
2 **a)** $39.3 \div 3$ **b)** $56.8 \div 8$
    c) $13.6 \div 4$ **d)** $15.6 \div 4$.

(ii) If dividing a decimal number by another decimal number first write the division as a fraction, then multiply the top number and the bottom number by 10 or 100 or ..... to get a whole number on the bottom and finally cancel or divide.

EXAMPLE 7

$0.9 \div 1.5 = \dfrac{0.9}{1.5} = \dfrac{9}{15} = \dfrac{3}{5} = 0.6$

EXERCISE 12.6A

Work out.

1 **a)** $0.42 \div 1.5$ **b)** $460 \div 0.2$ **c)** $1.43 \div 1.1$ **d)** $1.2 \div 1.5$
  e) $12.8 \div 3.2.$

2 **a)** $21.6 \div 2.4$ **b)** $16 \div 3.2$ **c)** $4 \div 0.8$ **d)** $7.4 \div 3.7.$

EXERCISE 12.6B

Work out.

1 **a)** $2.4 \div 0.6$ **b)** $17.5 \div 3.5$ **c)** $10.8 \div 1.2$ **d)** $12.4 \div 3.1.$
2 **a)** $1.8 \div 0.3$ **b)** $2.8 \div 0.7$ **c)** $4.2 \div 0.6$ **d)** $6.5 \div 1.3.$

Key ideas

- Adding a negative number is the same as subtracting a positive number.
- Subtracting a negative number is the same as adding a positive number.
- Line up decimal points when adding or subtracting decimal numbers.
- Multiplying or dividing decimal numbers by powers of 10 involves a 'movement' of the decimal point to the right, when multiplying, or the left when dividing.

13 Linear graphs I

You should already know

- how to plot and read points in the first quadrant, using (*x*, *y*) coordinates
- how to substitute in equations.

Points in all four quadrants

You can plot points in any of the four quadrants of the Cartesian plane, using positive and negative coordinates.

EXAMPLE 1

Make sure you get the sign correct.

The four points on the grid are A(2, 1), B(2, ⁻2), C(⁻2, ⁻3), D(⁻4, 3)

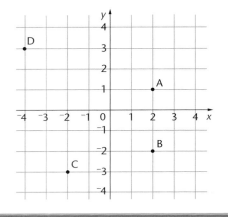

Exam tip

Always label the axes clearly.

Straight-line graphs

The simplest equations that give straight-line graphs are of the type $y = 3$ or $x = 2$.
Every point on the line $y = 3$ has y-coordinates 3.
Every point on the line $x = 2$ has x-coordinates 2.
The x-axis has the equation $y = 0$ and the y-axis has the equation $x = 0$.

So $y = 3$ is the equation of the line through 3 on the y-axis and is parallel to $y = 0$, and $x = 2$ is the equation of the line through 2 on the x-axis and is parallel to $x = 0$.

These are both drawn on the graph. Can you see the lines $y = {}^-2$ and $x = {}^-4$, which are also drawn on the graph?

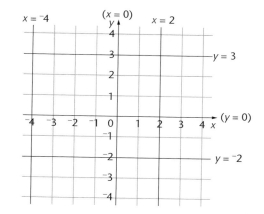

Lines of the type $y = mx + c$

The most common straight-line graphs you need to draw have equations of the form $y = mx + c$, for example $y = 3x + 4$, $y = 2x - 5$ and $y = {}^-3x + 2$.

You need only two points to draw a straight line, but you should work out three points to check. It is best to use two values that are as far apart as possible, and 0 if it is in the range.

EXAMPLE 2

Draw the graph of $y = 2x + 3$ for values of x from $^-3$ to $+3$.

You can use any three values of x from $^-3$ to $+3$. In this example the best values to use are $^-3$, 0 and $+3$.

When $x = {}^-3$, $y = 2 \times {}^-3 + 3$
$\qquad\qquad = {}^-6 + 3 = {}^-3$.

When $x = 0$, $y = 2 \times 0 + 3 = 0 + 3 = 3$.

When $x = +3$, $y = 2 \times 3 + 3 = 6 + 3 = 9$.

The graph needs to include x-values from $^-3$ to $+3$ and y-values from $^-3$ to $+9$.

Plot the points $(^-3, {}^-3)$, $(0, 3)$ and $(3, 9)$ and join them with a straight line. Label the line.

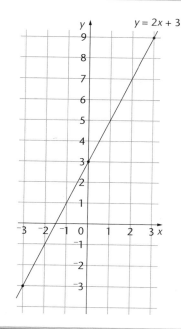

Exam tip

A common error is to say $2 \times 0 = 2$. Remember any number multiplied by zero is zero.
Write the values of x and y down clearly, as marks are often given for the correct calculation of the points even if they are plotted wrong.

EXAMPLE 3

Draw the graph of $y = -2x + 1$, for values of x from -4 to $+2$.

When $x = -4$, $y = -2 \times -4 + 1 = +8 + 1 = 9$.

When $x = 0$, $y = -2 \times 0 + 1 = 0 + 1 = 1$.

When $x = 2$, $y = -2 \times 2 + 1 = -4 + 1 = -3$.

In this case the y-values are from -3 to 9.

Plot the points $(-4, 9)$, $(0, 1)$ and $(2, -3)$ and join them with a straight line.

Notice that this line slopes the opposite way.

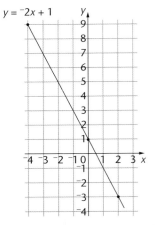

$y = -2x + 1$

Exam tip

Any line with equation $y = ax + b$
will cross the y-axis at $y = b$.

Exam tip

If axes are already drawn for you in a
question, check the scale carefully
before plotting points or reading values.

EXERCISE 13.1A

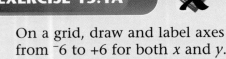

1 On a grid, draw and label axes
   from -6 to $+6$ for both x and y.

   Plot and label the points A(1, 3),
   B(-2, -6), C(-4, 3), D(-6, 3), E(2, -5),
   F(-2, 0), G(0, 5).

2 On a grid, draw and label axes
   from -6 to $+6$ for both x and y.
   Then draw the line for each of
   these equations.

   $x = 3$, $x = -3$, $y = 4$, $y = -5$.

3 Draw the graph of $y = 4x$, for $x = -3$
   to $+3$.

4 Draw the graph of $y = x + 3$, for
   $x = -3$ to $+3$.

5 Draw the graph of $y = 3x + 2$,
   for $x = -4$ to $+2$.

6 Draw the graph of $y = x - 3$,
   for $x = -2$ to $+4$.

7 Draw the graph of $y = 3x - 4$,
   for $x = -2$ to $+4$.

8 Draw the graph of $y = 4x - 2$,
   for $x = -3$ to $+3$.

9 Draw the graph of $y = -2x + 5$,
   for $x = -2$ to $+4$.

10 Draw the graph of $y = -3x - 4$,
   for $x = -4$ to $+2$.

EXERCISE 13.1B

1 On a grid, draw and label axes from $^-6$ to $+6$ for both x and y. Then draw the line for each of these equations.

$x = 5$, $x = ^-2$, $y = 2$, $y = ^-3$.

2 Draw the graph of $y = 3x$, for $x = ^-3$ to $+3$.

3 Draw the graph of $y = x + 6$, for $x = ^-4$ to $+2$.

4 Draw the graph of $y = 4x + 2$, for $x = ^-3$ to $+3$.

5 Draw the graph of $y = 3x + 5$, for $x = ^-4$ to $+2$.

6 Draw the graph of $y = x - 5$, for $x = ^-1$ to $+6$.

7 Draw the graph of $y = 2x - 5$, for $x = ^-1$ to $+5$.

8 Draw the graph of $y = ^-x + 1$, for $x = ^-3$ to $+3$.

9 Draw the graph of $y = ^-3x - 2$, for $x = ^-4$ to $+2$.

10 Draw the graph of $y = ^-2x - 4$, for $x = ^-4$ to $+2$.

Harder straight-line equations

Sometimes you could be asked to draw a straight line with an equation of the form $3x + 2y = 12$, where both the x-term and the y-term are on the same side of the equation. In this case, it is easier to work out the value of x when y is 0, and the value of y when x is 0.

Draw the line joining these two points. Then check that the coordinates of another point on the graph fit the equation.

EXAMPLE 4

Draw the graph of $4x + 3y = 12$.

When $x = 0$, $3y = 12$, $y = 4$.

When $y = 0$, $4x = 12$, $x = 3$.

The axes on the graph need to be labelled for x from 0 to 3, and for y from 0 to 4.

Plot the points $(0, 4)$ and $(3, 0)$ and join them with a straight line.

Check with a point on the line, such as $x = 1{\cdot}5$, $y = 2$.

$4x + 3y = 4 \times 1{\cdot}5 + 3 \times 2 = 6 + 6 = 12$, which is correct.

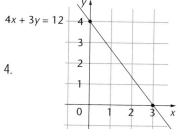

Other types of equation include $2y = 3x + 5$, where you need to find $2y$, then divide by 2.

Exam tip

A common error is to plot $(0, 4)$ at $(4, 0)$ and $(3, 0)$ at $(0, 3)$.

EXERCISE 13.2A

1 Draw the graph of $3x + 5y = 15$.
2 Draw the graph of $7x + 2y = 14$.
3 Draw the graph of $2y = 5x + 3$, for $x = {}^-3$ to $+3$.
4 Draw the graph of $2y = 3x - 5$, for $x = {}^-2$ to $+4$.
5 a) On the same grid, draw the graph of
    (i) $y = 8$ **(ii)** $y = 4x + 2$
    for $x = {}^-3$ to $+3$.

b) Write down the coordinates of the point where the two lines cross.

6 a) On the same grid, draw the graphs of $y = 2x + 3$ and $2x + y = 7$, for $x = 0$ to 5.
b) Write down the coordinates of the point where the two lines cross.

EXERCISE 13.2B

1 Draw the graph of $2x + 5y = 10$.
2 Draw the graph of $3x + 2y = 15$.
3 Draw the graph of $3y = 2x + 6$, for $x = {}^-3$ to $+3$.
4 Draw the graph of $2y = 5x - 8$, for $x = {}^-2$ to $+4$.
5 a) On the same grid, draw the graphs of
    (i) $y = x + 3$
    (ii) $y = 4x - 3$
        for $x = {}^-2$ to $+3$.
b) Write down the coordinates of the point where two lines cross.

6 a) On the same grid, draw the graphs of
$y = 3x - 2$ and $4x + y = 12$, for $x = 0$ to 5.
b) Write down the coordinates of the point where the two lines cross.

Key ideas

- Label axes clearly and check you understand the scales of graphs which have axes drawn for you.
- You only need two points to draw a straight line but check with a third point.

14 Linear graphs II

You should already know

● how to draw straight line graphs.

Story graphs

Some graphs tell a story – they show what happened in an event. To find out what is happening, first look at the labels on the axes. They tell you what the graph is about.

Look for important features on the graph. For instance, does it increase or decrease at a steady rate (a straight line) or is it curved?

Exam tip

When drawing a graph, don't forget to label the axes.

EXAMPLE 1

This graph shows the noise levels at a football stadium one afternoon. The boxes describe what may have caused the change in shape of the graph at certain points.

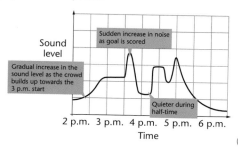

Exam tip

If you are asked to describe a story graph, try to include numerical information. For example, instead of 'stopped' write 'stopped at 10·14 p.m. for 6 minutes'.

EXAMPLE 2

John ran the first two miles to school at a speed of 8 m.p.h. He then waited 5 minutes for his friend. They walked the last mile to school together, taking 20 minutes.

The graph for this story has been started. Finish the graph. (The different colour on the graph shows where this has been done.)

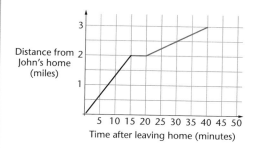

The first part of this graph is steeper than the last part. This shows that John went faster in the first 15 minutes than he did in the last 20 minutes. The flat part of the graph shows where John stayed in the same place for 5 minutes.

EXERCISE 14.1A

1 This graph shows the average monthly rainfall in Birmingham and Brussels.

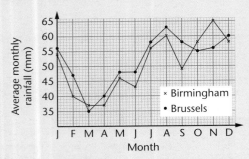

a) Which city is usually wetter in June?

b) What is the average rainfall in Birmingham in January?

c) For which month is there, on average, more rain in Birmingham than Brussels?

2 This line graph shows the average monthly temperature in Leeds.

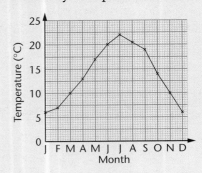

a) Which is the hottest month and what is its average temperature?

b) What is the average temperature in October?

c) After which month did the temperature start to decrease rapidly?

3 Amy drew this sketch graph to show what happened to the volume of water in her bath.

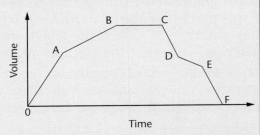

a) Both taps were 'on' at 0. What happened at A?

b) What happened at B?

c) What happened at C?

d) What happened between D and E?

EXERCISE 14.1B

1 The graph below shows how the depth of water in a water tank varies as the tank is filled.

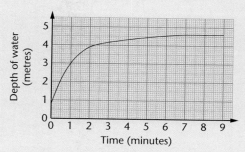

a) What is the depth of water to start with?

b) By how much does the water rise in the first minute?

c) At what time does the water reach the maximum depth? What is this depth?

2 This graph shows the temperature in a school during a 24 hour period.

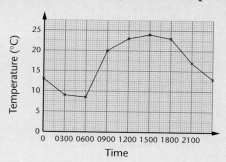

a) At what time did the heating switch on?

b) What was the temperature at 9 o'clock?

c) What is the difference between the lowest and highest temperature?

3 Some water was heated over a burner. Its temperature rose quickly at first and then more and more slowly. The burner was turned off and the temperature fell rapidly at first and then more and more slowly. Draw a sketch graph to show how the temperature changes.

Key ideas

- Label axes.
- Read scales carefully – they give a clue to what is happening.

15 Drawing triangles and other shapes

You should already know

- how to use simple scales
- how to use a pair of compasses and either a protractor or an angle measurer.

Drawing triangles

Given two sides and the included angle

The included angle is the angle between the two given sides.

EXAMPLE 1

Make an accurate drawing of the triangle sketched opposite.

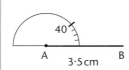

Draw the line AB and measure the 40° angle from A.

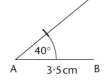

Draw a line from A through the point.

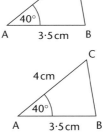

Mark the point C on this line, 4 cm from A. Join C to B.

EXERCISE 15.1A

1 Make accurate full-size drawings of these triangles.

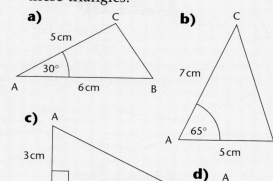

a)

5 cm C
30°
A   6 cm   B

b)

C
7 cm
A 65°
5 cm   B

c)

A
3 cm
C   6 cm   B

d)

A
4 cm 130°
C   6 cm   B

2 For each triangle in question 1, measure the unmarked side and the other two angles on your drawing.

3 For each triangle in question 1, find the perpendicular distance of C from AB on your drawing.

EXERCISE 15.1B

1 Draw these triangles accurately.

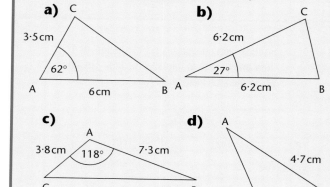

a)

C
3·5 cm
62°
A   6 cm   B

b)

C
6·2 cm
27°
A   6·2 cm   B

c)

A
3·8 cm 118° 7·3 cm
C   B

d)

A
4·7 cm
33°
C   2·8 cm   B

2 For each triangle in question 1, measure the unmarked side and the other two angles on your drawing.

3 For each triangle in question 1, draw a line through C parallel to AB on your drawing.

Given one side and two angles

EXAMPLE 2

In triangle PQR, PQ = 6 cm, angle RPQ = 35° and angle PQR = 29°.

Make an accurate drawing of triangle PQR.

Draw a sketch.

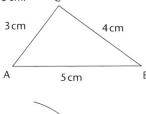

Draw the line PQ. Measure the angle of 35° at P and draw a line, any length but longer than PQ.

Measure the angle of 29° at Q. Draw a line to meet the previous line at R. This completes the triangle.

Exam tip

Draw a sketch of the triangle first. When an angle is written as three letters, the middle letter indicates the vertex (corner) of the angle.

Exam tip

If the two given angles were at R and Q, then angle P could be found by adding the two known angles and subtracting from 180°.

Given three sides

EXAMPLE 3

Make an accurate drawing of triangle ABC where AB = 5 cm, BC = 4 cm and AC = 3 cm.

Draw a sketch.

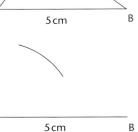

Draw the line AB. From A, with compasses set to a radius of 3 cm, draw an arc above the line.

From B, with compasses set to a radius of 4 cm, draw another arc to intersect the first. The point where the arcs meet is C.

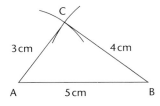

Given two sides and a non-included angle

A non-included angle is one which is not between the two known sides.

EXAMPLE 4

Make an accurate drawing of the triangle sketched opposite.

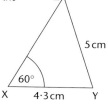

From Y, with compasses set to a radius of 5 cm, draw an arc to intersect the line. The point where the arc cuts the line is Z.

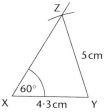

Draw the line XY. Measure the angle of 60° at X and draw a line of any length, but longer than XY.

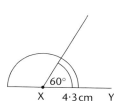

EXERCISE 15.2A

1 Make accurate full-size drawings of these triangles.

a)

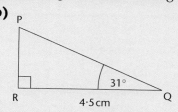

b)

c) Triangle PQR with PQ = 7 cm, angle RPQ = 41°, angle PQR = 53°

d) Triangle PQR with PR = 3·6 cm, angle PRQ = 126°, angle RPQ = 18°.

2 In each of the triangles in question 1, measure the unmarked sides on your drawing.

Exercise 15.2A cont'd

3 Draw these triangles accurately.

a)

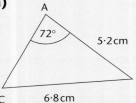

b)

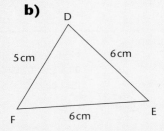

c) Triangle DEF with DE = 4·3 cm, EF = 7·2 cm, FD = 6·5 cm.

4 In each of the triangles in question 3, measure the angles on your drawing.

5 Draw these triangles accurately.

a)

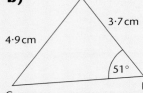

b)

(see figure)

EXERCISE 15.2B

1 Make accurate full-size drawings of these triangles.

a)

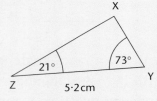

b)

(see figure)

c)

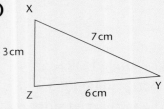

d)

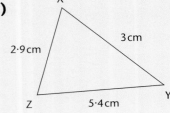

Exercise 15.2B cont'd

e) Triangle XYZ with XY = 4·9 cm, angle XYZ = 33°, angle YXZ = 66°

f) Triangle XYZ with XY = 8·3 cm YZ = 4·3 cm, ZX = 5·1 cm.

2 In each of the triangles in question 1, measure all the unknown lengths and angles on your drawings.

3 a) Draw the triangle ABC where AB = 5 cm, BC = 4 cm and angle BAC = 40°. Compare your drawing with your neighbour's drawing.

b) Draw triangle ABC where AB = 7 cm, AC = 6 cm and angle ABC = 50°. Compare your drawing with your neighbour's drawing.

There are several different methods to construct a regular polygon. Possibly the easiest method is to divide the angle at the centre of the circle, 360°, into a number of equal parts – the number of parts being equal to the number of sides of the polygon.

EXAMPLE 5

Construct a regular pentagon.

(i) Draw a circle with, for example, radius 5 cm.

(ii) A pentagon has 5 sides so divide the angle at the centre by 5 giving 360° ÷ 5 = 72°.

(iii) Measure, using a protractor, 72° angles around the centre, and join the centre to the circumference.

(iv) Join the points on the circumference.

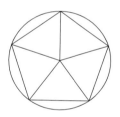

ACTIVITY 1

Now, using circles with the same radius construct a hexagon and an octagon.

Nets

This flat shape may be folded up to make a cube – it is called a net for a cube.

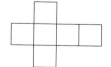

Chapter 15 *Drawing triangles and other shapes*

EXERCISE 15.3

1 Here are some more nets.
Which of these will fold into cubes?

a)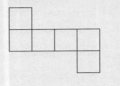

b)

c)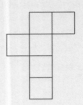

d)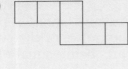

2 Which of the shapes a, b, or c could be the net of this cuboid?

a)

b)

c)

3 This is the net for a square based pyramid.

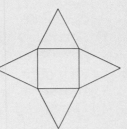

Is this a net for a different square based pyramid?

 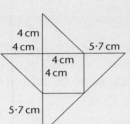

4 cm
4 cm
5·7 cm
4 cm
4 cm
5·7 cm

Exam tip

You could draw it accurately on squared paper and check by cutting the shape out and folding it up. You can draw the nets for pyramids by using the techniques described earlier in the chapter.

4 A 'Toblerone' box is a triangular prism.
Sketch the net of this triangular prism.

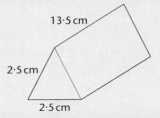

13·5 cm

2·5 cm

2·5 cm

Key ideas

- Accurate drawing must be accurate. Use a sharp pencil and check all measurements.
- Drawing a rough sketch first helps you to draw the sides and angles in the right place.
- Compasses are not just for drawing circles. Use them when you know the distance but not the direction.

C1 Revision exercise

1 The table below shows the heights and weights of a group of eight-people.

Height (cm)	140	132	164	169	157	185	176	143
Weight (kg)	38	29	60	57	55	73	68	41

Draw a scatter graph and see if there is any relationship between height and weight.

What would be the likely weight of someone 160 cm tall?

What would be the likely height of someone weighing 50 kg?

2 The scatter graph shows the English and history test scores of 7 boys and 7 girls. Compare the performance of the boys and the girls.

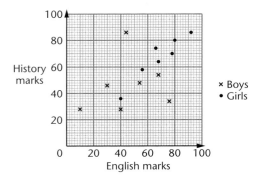

3 Nina thinks that the more she trains the lower the time it will take to swim a length.

State the type of correlation of which this is an example.

4 Two judges at an ice skating competition had to judge 8 skaters. The marks they gave, out of 15, are shown in the table.

Skater	A	B	C	D	E	F	G	H
Judge 1	7	9	8	12	13	11	9	6
Judge 2	8	7	6	9	10	10	6	6

a) Draw a scatter diagram to show the judges scores, with judge 1 along the horizontal axis.
b) Draw a line of best fit.
c) Judge 2 gave a ninth skater a score of 9.
Estimate the score that judge 1 would have given.

5 The heights, (in cm), and weights (in kg) of 20 members of a rugby club are given in the table.

height	180	182	184	181	179	178	186	188	192	189
weight	85	94	89	93	94	101	98	96	102	94

height	178	183	185	186	189	187	179	194	184	181
weight	99	97	98	94	89	102	91	96	94	98

a) What is the weight of the tallest player?
b) Draw a scatter graph to show this data.
c) Draw a line of best fit.
d) Estimate the weight of a player who is 1·98 m tall.

6 A botanist carried out an experiment to investigate how effective a new type of fertiliser was. She chose 10 plants and gave a different amount of fertiliser to each plant. The 10 plants were all initially the same height.
Sketch what you think the scatter diagram should look like, with the amount of fertiliser on the horizontal axis and the height of the plants on the vertical axis.

7 a) $1\frac{1}{2} \times 2\frac{2}{3}$ **b)** $2\frac{1}{5} \div 3\frac{2}{3}$
   c) $2\frac{3}{4} \times 1\frac{3}{4} \div 1\frac{3}{8}$.

8 a) £4·64 + £5·92
   b) £16·34 + £8·26
   c) £5·96 − £1·48
   d) £24·33 − £13·74

9 a) 2·46 + 1·70   **b)** 19·83 + 16·42
   c) 36·95 − 14·43   **d)** 134·2 − 99·9.

10 a) 8 × 7·6   **b)** 7 × 4·9
   c) 3 × 0·04   **d)** 8 × 0·9
   e) 5000 × 2·4   **f)** 6·99 × 400
   g) 1·88 × 3000.

11 a) 60·5 ÷ 5   **b)** 330·3 ÷ 3
   c) 3·244 ÷ 4   **d)** 85·5 ÷ 5
   e) 4 ÷ 0·8   **f)** 3·2 ÷ 0·4
   g) 3·9 ÷ 0·6   **h)** 5 ÷ 0·25.

12 On a piece of squared paper, draw a set of x- and y-axes, both labelled from ⁻6 to 6.

   a) Plot and label the points A(⁻3, 2), B(0, ⁻2), C(2, ⁻5), D(5, 0).
   b) Draw the lines $x = 3$, $x = ⁻5$, $y = ⁻3$, $y = 4$.

13 Draw a set of axes, labelling the x-axis from ⁻3 to 3 and the y-axis from ⁻10 to 12.

a) Draw the graph of $y = 3x$, for $x = ⁻3$ to 3.
b) Draw the graph of $x = ⁻2x + 5$.
c) Write down the coordinates of the point where two lines cross.

14 a) Draw the graph of $y = 2x − 4$, for $x = ⁻2$ to 5.
b) On the same grid, draw the graph of $5y + 2x = 10$.
c) Write down the coordinates of the point where the two lines cross.

15 Water was poured at a steady rate into this container until it was full.

Sketch a graph to show how the depth of water in the container changes with time.

16 'Beevis' car rental have three rates for the daily hire of their cars. For all three rates customers have to buy the petrol they use.

Rate A: 25p per mile
Rate B: £10 fixed charge plus 15p per mile
Rate C: £30 but with no mileage charge.

a) Find the cost of hiring a car for rate A if the distance travelled is:
   (i) 0 miles
   (ii) 30 miles
   (iii) 100 miles.

b) Find the cost of hiring a car for rate B if the distance travelled is:
   (i) 0 miles
   (ii) 30 miles
   (iii) 100 miles.

c) Find the cost of hiring a car for rate C if the distance travelled is:
   (i) 0 miles
   (ii) 30 miles
   (iii) 100 miles.

On the same axes draw three graph to show the cost of hiring a car for each rate, Use a scale of 20 miles for 2 cm on the *x*-axis and £10 for 2 cm on the *y*-axis.

d) Which is the cheapest rate if the distance to be travelled is:
   (i) 60 miles
   (ii) 120 miles
   (iii) 140 miles.

17 Here is the net of a solid.

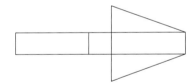

Which one of these solids could it be?

a)   **b)**

c)   **d)**

18 Which one of these sketches could be the net of a triangular prism?

a)

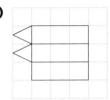

b)

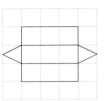

c)

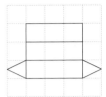

d)

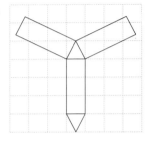

19 Describe the solid formed from each of these nets.

a)

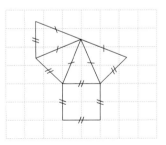

b)

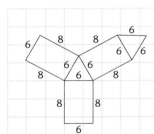

16 Areas

Area of a triangle

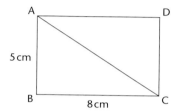

The area of the rectangle ABCD is $8 \times 5 = 40 \, cm^2$.

The area of the triangle ABC is half the area of the rectangle ABC and so is equal to $20 \, cm^2$.

So the area of the right-angled triangle $= \frac{1}{2} \times 8 \times 5 = 20 \, cm^2$.

Triangle PQR is not right-angled but has been split into two right-angled triangles.

Each right-angled triangle is half a rectangle.

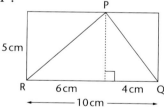

The area of the left-hand triangle is $\frac{1}{2} \times 6 \times 5 = 15 \, cm^2$.

The area of the right-hand triangle is $\frac{1}{2} \times 4 \times 5 = 10 \, cm^2$.

Therefore, the total area of the triangle is $10 + 15 = 25 \, cm^2$.

Notice that the area of triangle PQR is half the area of the large rectangle.

So the area of the triangle is $\frac{1}{2} \times 10 \times 5 = 25\,\text{cm}^2$.

This shows that whether the triangle is right-angled or not the area can be found by the formula:

area of triangle $= \frac{1}{2} \times$ base $\times$ perpendicular height or $A = \frac{1}{2} \times b \times h$

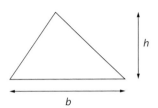

EXAMPLE 1

a) Find the area of this triangle.

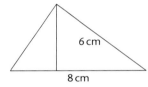

Answer:

a) Area $= \frac{1}{2} \times b \times h = \frac{1}{2} \times 8 \times 6 = 24\,\text{cm}^2$

b) If the area of this triangle is $20\,\text{cm}^2$, find the perpendicular height of this triangle.

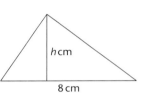

Answer:

b) Area $= \frac{1}{2} \times b \times h$.

$20 = \frac{1}{2} \times 8 \times h = 4h$

$\therefore h = 5\,\text{cm}$

Exam tip

Always use the perpendicular height of the triangle, never the slant height. So in the triangle above area $= \frac{1}{2} \times 6 \times 3 = 9\,\text{cm}^2$.

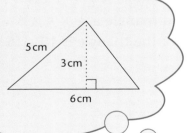

Remember that the units of area are always square units such as square centimetres or square metres, written cm² or m².

When using the formula, you can use any of the sides of the triangle as the base provided you use the perpendicular height that goes with it.

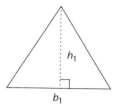 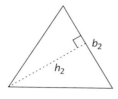 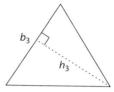

Area $A = \frac{1}{2} \times b_1 \times h_1$ Area $A = \frac{1}{2} \times b_2 \times h_2$ Area $A = \frac{1}{2} \times b_3 \times h_3$

EXERCISE 16.1A

1 Find the areas of these triangles.

a)
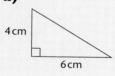
4 cm
6 cm

b)
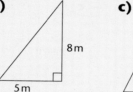
8 m
5 m

c)
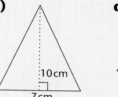
10 cm
7 cm

d)
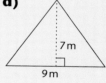
7 m
9 m

e)
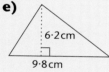
6·2 cm
9·8 cm

f)
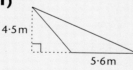
4·5 m
5·6 m

g)

5 m
3 m

h)
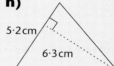
5·2 cm
6·3 cm

i)
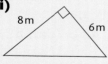
8 m
6 m

2 The vertices of a triangle are at A(2, 1), B(5, 1) and C(5, 7). Draw triangle ABC on squared paper and find its area.

3 The vertices of a triangle are at P(2, 2), Q(7, 2) and R(4, 6). Draw triangle PQR on squared paper and find its area.

Exercise 16.1A cont'd

4 The vertices of a triangle are at U(2, 6), V(9, 6) and W(5, 1). Draw triangle UVW on squared paper and find its area.

5 Using a ruler and compasses, draw an equilateral triangle with sides 5 cm. Measure its perpendicular height and find its area.

6 In triangle ABC, AB = 6 cm, BC = 8 cm and angle ABC = 40°. Draw the triangle accurately and find its area.

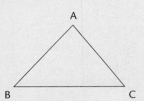

EXERCISE 16.1B

1 Find the areas of these triangles.

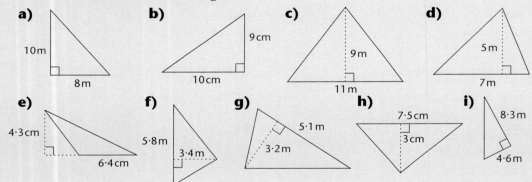

a) 10 m, 8 m

b) 9 cm, 10 cm

c) 9 m, 11 m

d) 5 m, 7 m

e) 4·3 cm, 6·4 cm

f) 5·8 m, 3·4 m

g) 5·1 m, 3·2 m

h) 7·5 cm, 3 cm

i) 8·3 m, 4·6 m

2 The vertices of a triangle are at A(2, 1), B(2, 7) and C(5, 3). Draw triangle ABC on squared paper and find its area.

3 The vertices of a triangle are at P(⁻2, 2), Q(3, 2) and R(5, 6). Draw triangle PQR on squared paper and find its area.

4 The vertices of a triangle are at U(⁻2, ⁻1), V(⁻2, 4) and W(3, 6). Draw triangle UVW on squared paper and find its area.

5 In triangle ABC, AB = 7 cm, BC = 10 cm and CA = 6 cm. Using ruler and compasses, draw the triangle accurately and find its area.

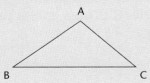

Exercise 16.1B cont'd

6 A triangle has an area of $12\,\text{cm}^2$ and a base of $4\,\text{cm}$. Find the perpendicular height associated with this base.

7 In triangle ABC, $AB = 6\,\text{cm}$, $BC = 8\,\text{cm}$ and $AC = 10\,\text{cm}$. Angle $ABC = 90°$.

a) Find the area of the triangle.
b) Find the perpendicular height BD.

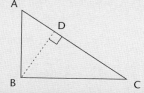

Area of a parallelogram

A parallelogram may be cut up and rearranged to form a rectangle or two congruent triangles.

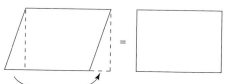

Area of a rectangle =
base × height.

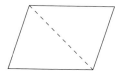

Area of a triangle =
$\frac{1}{2}$ × base × height.

Both these ways of splitting a parallelogram show:

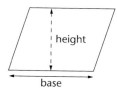

Area of a parallelogram =
base × height.

Exam tip

Make sure you use the perpendicular height and not the sloping edge when finding the area of a parallelogram.

EXAMPLE 1

Find the area of this parallelogram.

Area of a parallelogram = base × height

$$= 8 \cdot 3 \times 6 \cdot 2 \, \text{cm}^2$$

$$= 51 \cdot 46 \, \text{cm}^2$$

$$= 51 \cdot 5 \, \text{cm}^2 \text{ to three significant figures.}$$

6·2 cm

8·3 cm

Exam tip

Don't forget to give your final answer to a suitable degree of accuracy, but don't use rounded answers in your working.

EXERCISE 16.2A

1 Find the area of each of these parallelograms.

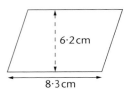

a)

5 cm

8 cm

b)

6 cm   7 cm

c)

7·5 cm

4 cm

2 Find the area of each of these parallelograms. The lengths are in centimetres.

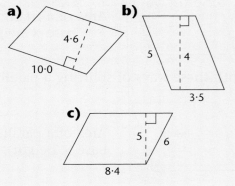

a)

4·6

10·0

b)

5   4

3·5

c)

5   6

8·4

Exercise 16.2A cont'd

3 Measure the base and height and calculate the area of each of these parallelograms.

a)

b)

c)

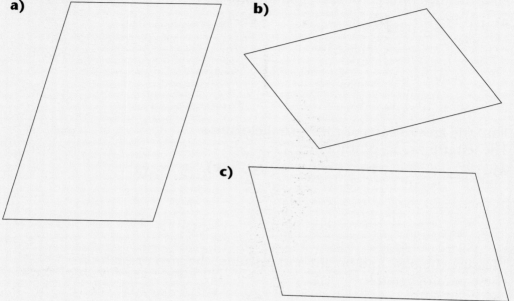

4 Find the values of x, y and z.

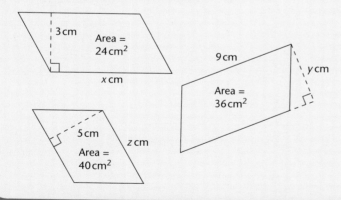

3 cm

Area = 24 cm²

x cm

9 cm

y cm

Area = 36 cm²

5 cm

z cm

Area = 40 cm²

EXERCISE 16.2B

1 Find the area of each of these parallelograms.

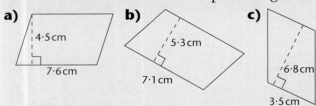

a) 4·5 cm, 7·6 cm

b) 5·3 cm, 7·1 cm

c) 6·8 cm, 3·5 cm

2 Find the area of each of these parallelograms.
The lengths are in centimetres.

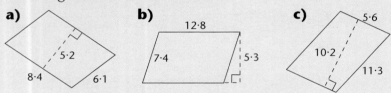

a) 5·2, 8·4, 6·1

b) 12·8, 7·4, 5·3

c) 5·6, 10·2, 11·3

3 Measure the base and height and calculate the area of each of
these parallelograms.

a)

b)

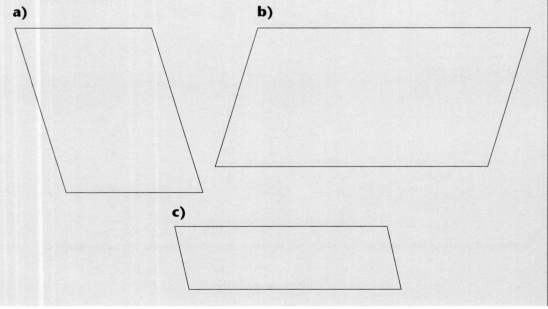

c)

Exercise 16.2B cont'd

4 Find the values of x, y and z.

a)

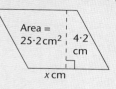

Area = 25·2 cm² 4·2 cm

x cm

b)

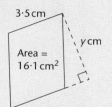

3·5 cm

y cm

Area = 16·1 cm²

c)

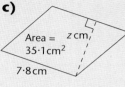

z cm

Area = 35·1 cm²

7·8 cm

Area of a trapezium

A trapezium has one pair of opposite sides parallel.
A trapezium can also be split into two triangles.

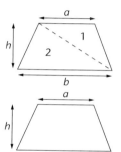

Area of triangle 1 = $\frac{1}{2} \times a \times h$

Area of triangle 2 = $\frac{1}{2} \times b \times h$

Area of trapezium = $\frac{1}{2} \times a \times h + \frac{1}{2} \times b \times h = \frac{1}{2} \times (a + b) \times h$

Area of a trapezium = $\frac{1}{2} \times (a + b)h$

In words, a useful formula to remember is:

Area of a trapezium = half the sum of the parallel sides × the height.

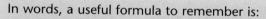

EXAMPLE 2

Calculate the area of this trapezium.

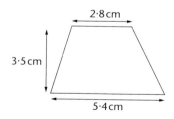

2·8 cm

3·5 cm

5·4 cm

Area of trapezium = $\frac{1}{2}(a + b)h$

$= \frac{1}{2}(2·8 + 5·4) \times 3·5 \text{ cm}^2$

$= 14·35 \text{ cm}^2$

$= 14·4 \text{ cm}^2$ to three significant figures.

Exam tip

When finding the area of a parallelogram or a trapezium, don't try to split the shape up. Instead, learn the area formulae and use them as it is quicker.

EXAMPLE 3

Find the area of this trapezium.

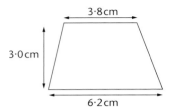

3·8 cm

3·0 cm

6·2 cm

Area of a trapezium = $\frac{1}{2}(a + b)h$

$\qquad = \frac{1}{2}(3·8 + 6·2) \times 3\,cm^2$

$\qquad = \frac{1}{2} \times 10 \times 3\,cm^2$

$\qquad = 15\,cm^2.$

Exam tip

When finding the area of a trapezium, an efficient method is to use the brackets function on your calculator. Without a calculator, work out the brackets first.

EXERCISE 16.3A

1 Find the area of each of these trapezia.

a)

7 cm
3 cm
13 cm

b)

3 cm
4 cm
5 cm

c)

4 cm
3 cm
6 cm

2 Find the area of each of these trapezia.

a)

7 cm   4 cm   3 cm

b)

6 cm
4 cm
2·5 cm

c)

3 cm
5 cm
9 cm

3 Measure the lengths you need and calculate the area of each of these trapezia.

a)

b)

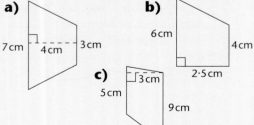

Exercise 16.3A cont'd

c)

5 A trapezia has height 4 cm and area 28 cm². One of its parallel sides is 5 cm long. How long is the other parallel side?

4 Find the values of *a*, *b* and *c* in these trapezia.

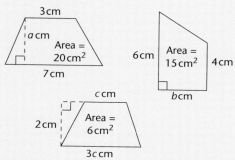

EXERCISE 16.3B

1 Find the area of each of these trapezia.

a)
7·1 cm
3·8 cm
9·5 cm

b)
2·8 cm
6·9 cm
4·7 cm

c)
3·6 cm
4·1 cm
5·9 cm

2 Find the area of each of these trapezia.

a)
3·7 cm
1·8 cm
2·1 cm

b)
1·1 cm
4·5 cm
5·7 cm

c)
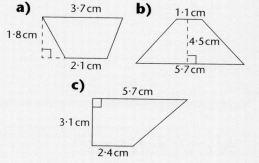

Exercise 16.3B cont'd

3 Measure the lengths you need and calculate the area of each of these trapezia.

a)

b)

c)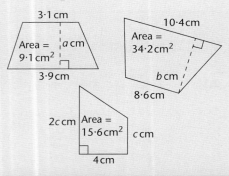

4 Find the values of a, b and c in these trapezia.

3·1 cm

Area = 9·1 cm² a cm

3·9 cm

10·4 cm

Area = 34·2 cm²

b cm

8·6 cm

$2c$ cm   Area = 15·6 cm² c cm

4 cm

5 A trapezium has height 6·6 cm and area 42·9 cm². One of its parallel sides is 5 cm long. How long is the other parallel side?

Key ideas

● You should know the formulas for the areas of triangle, rectangle and parallelogram.

● You need to know how to work out the area of a trapezium from the formula.

17 Volume

You should already know

- the common metric units for length, area and volume, i.e. cm, cm², cm³ or m, m², m³
- how to find the volume of a cuboid by counting cubes.

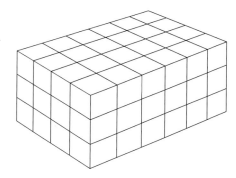

The diagram shows a cuboid made up of centimetre cubes.

On the top layer there are six rows of four cubes, so there are $6 \times 4 = 24$ centimetre cubes in the top layer.

There are three layers. So in all there are $3 \times 24 = 72$ cubes.

This means that the volume of the cuboid = 72 cubic centimetres or $72\,\text{cm}^3$.

The calculation for this was $6 \times 4 \times 3 = 72$, so the formula for the volume of a cuboid is:

Volume of a cuboid = V = length × width × height or $V = lwh$.

Exam tip

Make sure that the length, width and height are all in the same units. The units of volume are then the **cube** of those units e.g. cm³ or m³.

EXAMPLE 1

Calculate the volume of a cuboid with length 8·5 cm, width 6·4 cm and height 3·6 cm.

Volume = 8·5 × 6·4 × 3·6 = 195·84 cm^3.

Exam tip

When finding surface area it may help to sketch the net of the shape to check that you count all the sides of the shape.

EXAMPLE 3

A concrete path is laid which is 20 m long, 1·5 m wide and 10 cm thick. Calculate the volume of concrete used.

Thickness = 10 ÷ 100 = 0·1 m.

Volume = 20 × 1·5 × 0·1 = 3 m^3.

EXAMPLE 2

For the cuboid in the diagram, calculate

a) the volume   **b)** the total surface area.

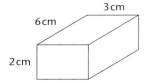

a) Volume = 6 × 3 × 2 = 36 cm^3.

b) Area of top = 6 × 3 = 18 cm^2

Area of side = 6 × 2 = 12 cm^2

Area of end = 2 × 3 = 6 cm^2

Total surface area of top + bottom + 2 sides + 2 ends

= 2 × 18 + 2 × 12 + 2 × 6

= 36 + 24 + 12

= 72 cm^2.

EXERCISE 17.1A

In all of these questions, make sure you state the units of your answers.

1 Find the volumes of the following cuboids:

  a) L = 4 cm, W = 3 cm, H = 2 cm

  b) L = 13 cm, W = 6 cm, H = 8 cm

  c) L = 10 cm, W = 8 cm, H = 4·5 cm

2 A classroom is in the shape of a cuboid. It has length 6·8 m, width 5·3 m and height 2·8 m. Calculate the volume of the room.

3 Measure the length, width and thickness of this text book. What is the volume of the text book?

4 For the cuboid shown, calculate:

  a) the volume

  b) the total surface area.

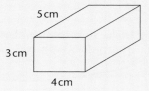

Exercise 17.1A cont'd

5 A rectangular fish pond is 2·5 m long and 1·4 m wide. The depth of the water is 40 cm. Calculate the volume of water in cm^3. How many litres is this? (1 litre = 1000 cm^3)

6 The diagram shows the net of a shoe box. Calculate the volume of the shoe box.

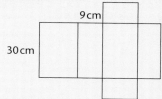

EXERCISE 17.1B

In all of these questions, make sure you state the units of your answer.

1 Find the volumes of the following cuboids:
   a) $L = 14$ cm, $W = 8$ cm, $H = 8$ cm
   b) $L = 6·5$ cm, $W = 4$ cm, $H = 3·5$ cm
   c) $L = 9$ cm, $W = 7·5$ cm, $H = 4·3$ cm

2 Find the volume of a cube of side 20 cm.

3 A midi stereo system is 35 cm wide, 27 cm in depth (front to back) and 39 cm high. Assuming the system to be a cuboid, find its volume.

4 A freezer is 0·6 m wide, 0·6 m deep and 1·4 m high.
   a) Assuming the freezer to be a cuboid, find its volume.
   b) About 35% of the volume can be used for storing food. Calculate the volume of food that can be stored.

5 Measure the length, width and height of your bedroom. (If it is not approximately a cuboid choose another room in your home.) Calculate the volume of your bedroom.

6 Alan is laying concrete to make a garden path 18 metres long and 0·8 metres wide. If the concrete is to be 10 cm deep, how many cubic metres of concrete should he order?

Exercise 17.1B cont'd

7 For the cuboid shown, calculate:
  a) the volume of the cuboid
  b) the total surface area of
  the cuboid.

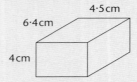

8 **a)** A fish tank is 80 cm long, 45 cm
  wide and 40 cm high. Ashraf
  filled it with water to a depth
  of 35 cm. Find the volume of
  water he used.

  b) Ashraf used a 2 litre jug to fill
  the tank with water. How
  many times did he fill the jug?
  (1 litre = 1000 cm^3).

9 **(i)**

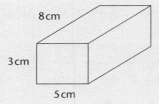

(ii)

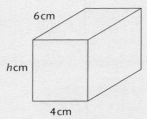

The two cuboids have the same
volume.

  a) Find h.

  b) Which cuboid has the smaller
  surface area? Show your
  calculations.

Key ideas

● The volume of a cuboid = length $\times$ width $\times$ height = *lwh*.

● The units of volume are cubic units such as cubic metres or cubic centimetres
  written as m^3 or cm^3.

18 Grouping data

You should already know

- how to find the median, mean, mode and range of a set of data
- how to use them to compare data.

Stem and leaf diagrams

Here are the marks gained by 30 students in an examination:

63	58	61	52	59	65	69	75	70	54
57	63	76	81	64	68	59	40	65	74
80	44	47	53	70	81	68	49	57	61

A different way of showing these data is to make a 'stem and leaf' diagram like this:

1 Write the tens figures in the left hand column of a diagram. These are the 'stems'

```
4
5
6
7
8
```

2 Go through the marks in turn and put in the units figures of each mark in the proper row. These are the 'leaves'

first 63
```
4
5
6 3
7
8
```

then 58
```
4
5 8
6 3
7
8
```

then 61
```
4
5 8
6 3 1
7
8
```

3 When all the marks are entered the diagram will look like this.

```
4 0 4 7 9
5 8 2 9 4 7 9 3 7
6 3 1 5 9 3 4 8 5 8 1
7 5 0 6 4 0
8 1 0 1
```

You must also add a key.

4 Finally re-write the diagram so the units figures in each row are in size order, with the smallest first.

```
4 0 4 7 9
5 2 3 4 7 7 8 9 9
6 1 1 3 3 4 5 5 8 8 9
7 0 0 4 5 6
8 0 1 1
```

Key: 5|2 = 52

This is a finished 'stem and leaf' table. It is a sort of frequency chart and allows you to read off certain information, for example:

- the modal group, (the one with the highest frequency), is the 60–69 group
- there are 30 results so the median result is mid-way between the 15th and the 16th result. Starting at the first result, 40, and counting on 15 results gives 63, the 16th result is also 63 so the median result will be 63.

EXERCISE 18.1A

1 These are the weights, in kilograms, of 25 new born babies.

2·6   2·9   3·2   2·5   3·1   1·9   3·5   3·9   4·0   2·8   4·1   1·7
3·8   2·6   3·1   2·4   4·1   2·6   4·2   3·6   2·9   2·8   2·7   3·3
3·8

a) Copy and complete this stem and leaf table:

```
1 |
2 |
3 |
4 |
```

Key: 1|7 = 1.7

b) Use your table to find the median weight.

2 A group of pupils took two mathematics tests. The stem and leaf tables of the marks for the tests are:

Test 1								Test 2							
2	3	4	5					2	0	1	2	6			
3	1	3	6	8				3	1	2	2	4	5	7	8
4	0	2	5	6	7			4	2	2	5	5	6	8	8 9
5	2	3	5	7				5	0	2	2	6	7	8	
6	1	3	5	5	6	9		6	1	3	4	4	9	9	
7	0	1	3	4	4	6 8		7	0	2	3	5	7		
8	0	2	3	3	6	6 7		8	3	7	9				
9	1	2	2	4	5			9	1	5					

Key: 3|3 = 33

a) Which test appeared to be harder?
b) Find the median mark for each test.

Exercise 18.1A cont'd

3 A scientist measured the lengths of a type of fish caught in two different rivers.

Here are their lengths, in cm:

River A

38	47	43	51	45	33	62	57
36	40	49	66	55	49	45	31
40	44	57	58	35	52	73	39
38	69	46	55				

River B

48	52	54	37	42	65	70	49
61	50	54	45	61	72	74	64
56	38	65	69	71	67	71	70
68							

a) Make a stem and leaf table to show the lengths of fish in each river.
b) Find the median length for the fish caugh at each site.

EXERCISE 18.1B

1 The speeds of 30 cars, in mph, measured in a city street are given below:

41	15	26	14	28	22	27	18	21	32	43
37	30	25	18	25	29	34	28	30	25	52
36	9	21	25	16	29	32	19			

a) Copy and complete the stem and leaf table to show this data.

```
0 |
1 |
2 |
3 |
4 |
5 |
```

b) Use your table to find the median speed.
c) If the speed limit were 30 mph what percentage of carse were breaking this limit?

2 A TV repair company monitored the time, in hours, it took their engineers to visit 20 homes and repair or replace televisions and videos.

0·9	1·0	2·1	2·4	0·7	1·1	0·9	0·6	0·4	0·3
1·2	1·6	0·6	0·3	0·7	1·4	1·0	0·8	0·7	0·9

Chapter 18 *Grouping data*

Exercise 18.1B cont'd

a) Copy and complete the stem and leaf table to show this data

```
0·|
1·|
2·|
```

b) Use your table to find the median time.

3 Rakhi's birthday is December 12th. She decided to look in a newspaper and see how old other people were who shared the same birthday. Her results were as follows:

56	74	75	56	67	71	48	30	45	58
60	62	21	24	36	38	51	56	43	22
18	32	44	40	21	18	50	40	30	50

a) Show this data in a stem and leaf table, choosing a suitable interval.

b) What is the median age?

4 Mr. Smith recorded the marks his students got on a history test. These are the marks they scored:

| 34 | 56 | 54 | 76 | 84 | 48 | 32 | 18 | 43 | 66 | 50 |
| 67 | 52 | 43 | 44 | 76 | 88 | 35 | 44 | 60 | 39 | 43 |

a) Show this data in a stem and leaf table, choosing a suitable interval.

b) What is the median mark?

Grouping discrete data

When working with large amounts of data, it is often easier to see the pattern of the data if they are grouped. For example, this is a list of goals scored in 20 matches.

| 1 | 1 | 3 | 2 | 0 | 0 | 1 | 4 | 0 | 2 |
| 2 | 0 | 6 | 3 | 4 | 1 | 1 | 3 | 2 | 1 |

Number of goals	Frequency
0	4
1	6
2	4
3	3
4	2
5	0
6	1

Mode

From the table, it is easy to identify the number of goals with the greatest frequency. That is the mode of the data. Here the mode is 1 goal.

Mean

The table can also be used to calculate the mean.

There are: four matches with 0 goals $4 \times 0 = 0$ goals
            six matches with 1 goal $6 \times 1 = 6$ goals
            four matches with 2 goals $4 \times 2 = 8$ goals

and so on.

To find the total number of goals scored altogether, multiply each number of goals by its frequency and then add the results.

Then dividing by the total number of matches (20) gives the mean.

The working for this is shown in the table.

Mean = 37 ÷ 20 = 1·85 goals.

In this example you have been given the original data. Check the answer by adding up the list of goals scored and dividing them by 20!

Number of goals	Frequency	Number of goals × frequency
0	4	0
1	6	6
2	4	8
3	3	9
4	2	8
5	0	0
6	1	6
Totals	20	37

Exam tip

When given a table of grouped data, add an extra column if necessary to help you work out the values multiplied by their frequencies.

Chapter 18 Grouping data

EXAMPLE 1

Work out the mean, mode and range for the number of children in the houses in Berry Road, listed in this table.

Number of children (c)	Frequency (number of houses)	c × frequency
0	6	0
1	4	4
2	5	10
3	7	21
4	1	4
5	2	10
Totals	25	49

Mean = 49 ÷ 25 = 1·96 children

Mode = 3 children

Range = 5 − 0 = 5 children

EXERCISE 18.2A

1 Mrs. Smith counts the number of spelling mistakes on each page of John's essay. Here are her findings.

20 19 6 5 14 17 16 13 16 11 8 15

a) Make a frequency table for the results.

b) What is the modal number of errors?

c) Calculate the mean number of errors.

2 Tom is the manager of a garage. He records the numbers of one make of car that are serviced each week day during a 4 week period.

8 6 5 2 8 10 8 9 6 5

5 9 10 8 6 5 8 9 10 15

a) Write down the modal number of cars serviced.

b) Find the median number of cars serviced.

c) Find the mean number of cars serviced per day.

3 Bagthorpe United football team record the number of goals it scored each week for the 200–2001 football season.

Number of goals	0	1	2	3
Frequency	7	10	9	4

a) How many matches were played?

b) What was the total number of goals scored?

c) What was the mean number of goals scored?

d) What was the median number of goals scored?

Exercise 18.2A cont'd

4 Christine decides to count the number of items of 'junk' mail she receives through the post each day during October.

Number of items	0	1	2	3	4	5	6
Number of days	5	4	6	7	2	0	1

a) What was the mode of the number of items?

b) What was the mean of the number of items?

EXERCISE 18.2B

1 A management company does a survey of the number of women working as managers in 15 different firms in a city.

Firm	A	B	C	D	E	F	G
Number of women	15	9	8	3	7	10	3

Firm	H	I	J	K	L	M	N	O
Number of women	10	3	27	21	4	34	7	8

Find:
  a) the modal number of women
  b) the mean number of women

2 This table shows the number of minutes late that the pupils in form 7B arrived for registration one Monday morning.

minutes late	1	2	3	4	5	6	7	8	9	10
frequency	0	1	5	7	11	2	1	0	1	0

There are 28 pupils in the class.
  a) How many pupils were on time?
  b) What was the modal number of minutes late?
  c) What is the mean number of minutes late?

3 Jenny works at a garden centre. She sows 50 seeds in each of ten seed trays and counts the number of plants that grow.

Tray	1	2	3	4	5	6	7	8	9	10
number of plants	48	43	45	46	42	49	46	47	50	41

  a) What is the modal number of plants?
  b) What is the mean number of plants?
  c) How many seeds did not germinate and grow into plants?

Representing continuous data

When data involve measurements they are always grouped, even if they don't look like it. For instance, a length L given as $18\,cm$ to the nearest centimetre means $17{\cdot}5 \leqslant L < 18{\cdot}5$. Any length between these values will count as $18\,cm$. Often, however, the groups are larger, to make handling the data easier. For example, when recording the heights, $h\,cm$, of 100 students in year 11, groups such as $180 \leqslant h < 185$ may be used.

Bar graph

Continuous data may be represented on a bar graph, using proper scales on both axes. Where the groups are not of the same width, the graph is called a **histogram**, and the **area** of each bar represents the frequency. In this chapter, however, only bars of equal width are considered, and the height of the bars represents the frequency, as with bar charts you have studied before.

EXAMPLE 2

Draw a bar graph to represent this information about the heights of students in year 11 at Sandish School.

Height ($h\,cm$)	Frequency
$155 \leqslant h < 160$	2
$160 \leqslant h < 165$	6
$165 \leqslant h < 170$	18
$170 \leqslant h < 175$	25
$175 \leqslant h < 180$	9
$180 \leqslant h < 185$	4
$185 \leqslant h < 190$	1

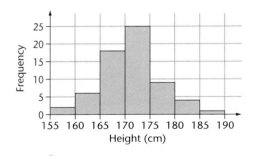

Exam tip

Check that you have labelled both scales carefully and that the boundaries of your bars match the boundaries of the groups.

Frequency polygons

A frequency polygon may also be used to represent the data. However, in this case, only one point is used to represent each group. The midpoint value of each group is chosen, as it is an average value for the group.

Exam tip

To work out the midpoint of a group, add together its boundary values and divide by 2.

EXAMPLE 3

Show the heights of the year 11 students in Sandish School in a frequency polygon.

The midpoint of the $155 \leqslant h < 160$ group is $\dfrac{155 + 160}{2} = 157{\cdot}5$.

So the points are plotted at h-values of 157·5, 162·5, 167·5 and so on.

Height (h cm)	Frequency	Midpoint
$155 \leqslant h < 160$	2	157·5
$160 \leqslant h < 165$	6	162·5
$165 \leqslant h < 170$	18	167·5
$170 \leqslant h < 175$	25	172·5
$175 \leqslant h < 180$	9	177·5
$180 \leqslant h < 185$	4	182·5
$185 \leqslant h < 190$	1	187·5

Exam tip

It can be helpful to add a column to the frequency table, like this.

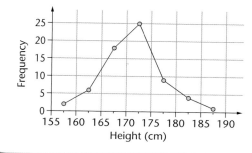

Calculating with grouped continuous data

Finding the mean

When working out the mean of grouped data in a table, you do not know the exact value for each item of data, so again the midpoint value is chosen to represent each group, and this is used to calculate an estimate of the mean. The midpoint is multiplied by the frequency of the group, as in calculating the mean of grouped discrete data.

EXAMPLE 4

Calculate an estimate of the mean height of the students in year 11 at Sandish School.

Height (h cm)	Frequency	Midpoint	Midpoint × frequency
$155 \leqslant h < 160$	2	157·5	315
$160 \leqslant h < 165$	6	162·5	975
$165 \leqslant h < 170$	18	167·5	3015
$170 \leqslant h < 175$	25	172·5	4312·5
$175 \leqslant h < 180$	9	177·5	1597·5
$180 \leqslant h < 185$	4	182·5	730
$185 \leqslant h < 190$	1	187·5	187·5
Totals	65		11 132·5

Mean = 11 123·5 ÷ 65 = 171·3 cm, correct to one decimal place.

Exam tip

Add two columns to the frequency table to help you work out the mean – one column for the midpoints of each group and one for the midpoints multiplied by their frequencies.

Mode and range of grouped continuous data

The **modal class** may be found when data are given as a table, frequency polygon or bar graph. It is the class with the highest frequency.

Exam tip

When stating the modal class, take care to give its boundaries accurately.

The **range** cannot be stated accurately from grouped data. For instance, the height of the tallest student in the example might be 189·7 cm or 185·0 cm. As with the mean, the midpoints of the groups are used to estimate the range.

EXAMPLE 5

For the heights of the year 11 students in Sandish School,

a) state the modal class

b) estimate the range.

a) The modal class is the one with the largest frequency, which is 170 cm ? height ? 170–175 cm.

b) An estimate of the range is found by finding the difference between the midpoint values of the top and bottom groups in the table. This gives range = 187·5 − 157·5 = 30 cm.

EXERCISE 18.3A

1 State the midpoints of these intervals.
   a) 20 m < length ≤ 30 cm
   b) 3 litres ≤ volume ≤ 3·5 litres
   c) 45 kg ≤ mass < 50 kg
   d) 3 mm < length ≤ 5 mm

2 Calculate an estimate of the mean age of the following group of men.

Age (years)	20≤age<30	30≤age<40	40≤age<50	50≤age<60
Frequency	3	8	5	2

Age (years)	60≤age<70	70≤age<80	80≤age<90
Frequency	4	7	5

3 **a)** Calculate an estimate of the mean of these heights.

Height (cm)	50–60	60–70	70–80	80–90	90–100
Frequency	15	23	38	17	7

   b) Draw a frequency polygon to represent this distribution.

Exercise 18.3A cont'd

4 Calculate an estimate of the mean of these lengths.

Length (y cm)	1·0–1·2	1·2–1·4	1·4–1·6	1·6–1·8	1·8–2·0
Frequency	2	7	13	5	3

5 Draw a bar graph to show this information.

Length (y cm)	Frequency
$10 \leqslant y < 20$	2
$20 \leqslant y < 30$	6
$30 \leqslant y < 40$	9
$40 \leqslant y < 50$	5
$50 \leqslant y < 60$	3

6 For the data in question 5:
   a) state the modal class
   b) calculate an estimate of the mean.

7 Draw a frequency polygon to show these data.

Mass of tomato (t g)	Frequency
$35 < t \leqslant 40$	7
$40 < t \leqslant 45$	13
$45 < t \leqslant 50$	20
$50 < t \leqslant 55$	16
$55 < t \leqslant 60$	4

8 For the data in question 7:
   a) estimate the range
   b) calculate an estimate of the mean.

9 This bar chart shows the ages of a group of men travelling on the 08:00 train to London one Friday morning.
   a) Make a frequency table for the data
   b) Estimate the range of the ages
   c) Estimate the mean of the ages

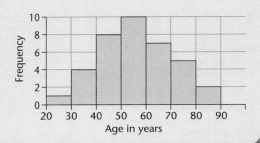

EXERCISE 18.3B

1 State the midpoints of these intervals.
   a) $25\,cm \leqslant length < 35\,cm$
   b) $0{\cdot}1\,cm \leqslant length < 0{\cdot}2\,cm$
   c) $500\,kg \leqslant mass < 520\,kg$
   d) $10\,ml \leqslant volume < 14\,ml$
   e) $0\,s \leqslant time < 5\,s$

2 Calculate an estimate of the mean of these heights

height, h, (cm)	$0 \leqslant h < 20$	$20 \leqslant h < 40$	$40 \leqslant h < 60$	$60 \leqslant h < 80$	$80 \leqslant h < 100$
Frequency	6	7	3	5	9

3 a) Calculate an estimate of the mean of these heights.

Height (m)	0–2	2–4	4–6	6–8	8–10
Frequency	12	26	34	23	5

   b) Draw a frequency polygon to represent this distribution.

4 Calculate an estimate of the mean of these lengths.

Length (cm)	3·0–3·2	3·2–3·4	3·4–3·6	3·6–3·8	3·8–4·0
Frequency	3	8	11	5	3

5 Draw a bar graph to show this information.

6 For the data in question 5:
   a) state the modal class
   b) calculate an estimate of the mean.

Mass (wkg)	Frequency
$30 \leqslant w < 40$	5
$40 \leqslant w < 50$	8
$50 \leqslant w < 60$	2
$60 \leqslant w < 70$	4
$70 \leqslant w < 80$	1

7 Draw a frequency polygon to show these data.

8 For the data in question 7:
   a) estimate the range
   b) calculate an estimate of the mean.

Length (xcm)	Frequency
$0 < x \leqslant 5$	8
$5 < x \leqslant 10$	6
$10 < x \leqslant 15$	2
$15 < x \leqslant 20$	5
$20 < x \leqslant 25$	1

Exercise 18.3B cont'd

9 This bar chart shows the playing times for a selection of CDs.
a) Make a frequency table for the data
b) Calculate an estimate of the mean playing time

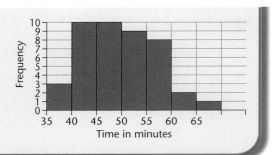

Mean, median or mode?

Each of these terms may be called an average. Sometimes, you need to decide which of them is the best to use in a given situation.

The mean is what most people think of as the average, but it may be 'fairer' to use the mode or median.

Look at these annual wages, for example.
The modal group is £15 000–20 000.

Annual wage (£)	Number of employees
10 000–15 000	2
15 000–20 000	18
20 000–25 000	12
25 000–30 000	4
30 000–35 000	0
35 000–40 000	2

Calculating the mean gives:

Midpoint	Frequency	Midpoint × frequency
12 500	2	25000
17 500	18	315000
22 500	12	270000
27 500	4	110000
32 500	0	0
37 500	2	75000
Totals	38	795000

Mean = 795 000 ÷ 38 = 20 921·0526 = £20 900 to the nearest £100.

Which gives a better idea of the average here, the mean or the mode? It depends on the purpose for which you want to use the average. If you were arguing for a pay rise you would probably use the mode. If you were the management, you would be more likely to use the mean.

If a distribution is quite symmetrical, there is not much difference between the mean, median and mode. If the distribution is weighted to one side (or **skewed**), then it matters much more which is chosen. In this case, the mean may not give the most representative average. Be prepared to give a reason why you have chosen to use a particular average.

A frequency polygon is constructed by joining the midpoint of the top of the bars, in order, in a frequency diagram.

The mode is the number or value which occurs most often.

The median is the middle value or, if there is an even number of values, it is halfway between the middle two values.

The mean is the usual average and is found by adding up all the values and dividing by the number of values.

The range is the difference between the smallest and largest values.

19 Enlargement

Enlargements

If you draw an enlargement of a shape, the image is the same shape as the original, but it is larger or smaller. You may just be asked to draw a '3 times enlargement' of a simple shape, without being given any other information. In this case copy the shape, making each side three times as long as the original one.

In the example, each side in the image is twice the length of the corresponding side in the original. The enlargement and the original are not congruent because they are different sizes, but they are similar.

More usually, you will be asked to draw an enlargement from a given point. In this case the enlargement has to be the correct size and also be in the correct position. The given points is the **centre of enlargement**.

For a '2 times enlargement', each point in the image must be twice as far from the centre as the corresponding point in the original is.

EXAMPLE 1

Make a '2 times enlargement' of this shape.

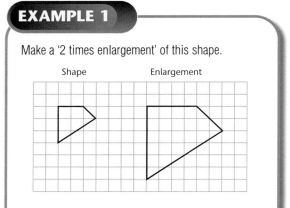

Shape   Enlargement

The enlargement can be drawn anywhere, but make sure it will fit on the grid.

If the centre is on the original shape, that point will not move and the enlargement will contain the original shape.

The number used to multiply the lengths for the enlargement is called the scale factor and is sometimes called *k*.

You may be asked to describe an enlargement. To do this you must give the scale factor and the centre of enlargement.

EXAMPLE 2

Enlarge this shape by scale factor 3 with the origin as centre.

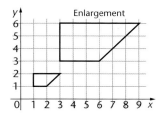

Scale factor 3 means this is a '3 times enlargement', so each point in the image is three times as far from the original as the corresponding point in the original shape is.

The image of the point (1, 1) is at (3, 3). The image of the point (2, 1) is at (6, 3) and so on.

EXERCISE 19.1A

Answer these questions on squared paper. In each case copy the original diagram.

1 Draw a '4 times enlargement' of this shape.

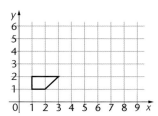

2 Draw a '3 times enlargement' of this shape.

3 Enlarge this shape by scale factor 2, centre the origin.

4 Enlarge this shape by scale factor 2, centre the point A.

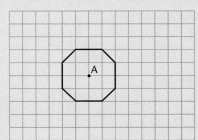

Exercise 19.1A cont'd

5 Enlarge this shape by a scale factor 3, centre the point A.

Note that the enlargement includes the first triangle and A does not move.

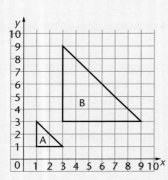

6 Enlarge this shape by scale factor 2, centre the point (2, 1).

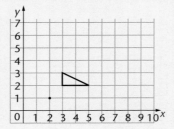

7

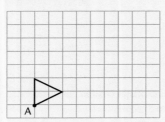

Describe the transformation that maps A on to B.

8

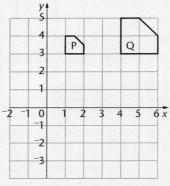

Describe the transformation that maps P on to Q.

EXERCISE 19.1B

Use squared paper to answer these questions. In each case copy the original diagram.

1 Draw a '2 times enlargement' of this shape.

2 Draw a '3 times enlargement' of this shape.

Exercise 19.1B cont'd

3 Enlarge this shape by scale factor 3, centre the origin.

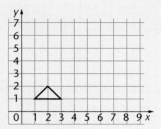

4 Enlarge this shape by scale factor 2, centre the point A.

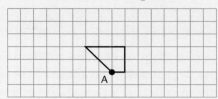

5 Enlarge this shape by scale factor 3, centre the point A.

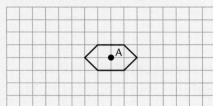

6 Enlarge this shape by scale factor 2, centre the point (2, 2).

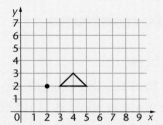

7

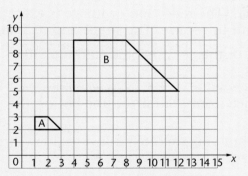

Describe the transformation that maps A on to B.

8

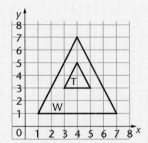

Describe the transformation that maps T on to W.

When a shape is enlarged by, e.g. a scale factor of 2, from a given centre, each point on the enlargement will be 2 times as far away from the centre as the corresponding point on the original shape is.

A shape and its enlargement are similar.

20 Transformations

Transformations

Reflections

You may find tracing paper helpful with some of these transformations.

You have already looked at reflections in a vertical line, but you need to be able to reflect a shape in any mirror line.

When a shape is reflected it is 'turned over'. The reflection or image is exactly the same size and shape as the original, but in reverse. The shape and its image are congruent. For example the reflection of this ▷ shape would look like this ◁ .

Corresponding points are the same distance from the mirror line but on the opposite side.

Reflections can be drawn on plain paper, but you are often asked to draw them on squared paper.

In this diagram the shape has been reflected in the line PQ.

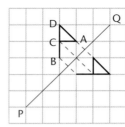

This means:

- the image of point A is half a square diagonally on the other side of the line PQ
- the image of B is half a square diagonally on the other side
- the image of C is a full square diagonally on the other side
- the image of D is one and a half squares diagonally on the other side.

Instead of counting squares you could trace the shape and turn the tracing paper over on the mirror line.

EXAMPLE 1

Reflect this shape in the line AB.

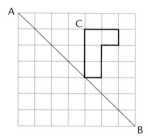 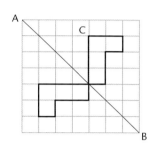

Exam tip

When you have drawn a reflection in a sloping line, check it by turning the page so the reflection line is vertical. Then you can easily see if it has been reflected correctly.

One point is on the line, so its image stays there. The images of the other points are each the same distance from the line as the original point, but on the opposite side. For example, point C and its image are both one and a half squares from the mirror line.

Rotations

When a shape is rotated it is turned relative to a fixed point, to a different angle. The shape is still the same way round, and the original shape and its image are

congruent. For example this shape  could turn to any of these positions:

 .

To rotate a shape you need to know three things:

- the angle of rotation
- the direction of rotation
- the centre of rotation.

Remember that rotations are anticlockwise, unless you are told otherwise.

Chapter 20 *Transformations*

This chapter covers:

- quarter-turns (90° anticlockwise)
- half-turns (180°)
- three-quarter turns (90° clockwise)

about the centre of the shape or the origin.

In this diagram the shape P has been rotated through a quarter-turn (90° anticlockwise) about point O.

You can do this by counting squares, or by tracing the shape and turning it, or by using a pair of compasses with the point at O and drawing quarter circles from each point.

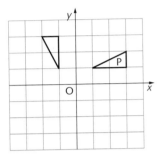

EXAMPLE 2

Rotate the shape through a three-quarter-turn about O.

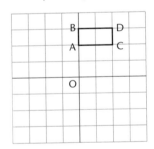 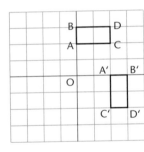

> **Exam tip**
>
> When you have drawn the rotation, turn the page through the correct angle to check it looks like the original.

You can think of a three-quarter-turn (anticlockwise) as a quarter-turn clockwise.

You can draw the rotation by counting squares.

A is 2 squares above O, so its image A' is 2 squares to the right.

B is 3 squares above O, so its image B' is 3 squares to the right.

C is 2 squares above and 2 to the right of O, so its image C' is 2 to the right and 2 below O.

D is 3 squares above and 2 to the right of O, so its image D' is 3 to the right and 2 below O.

Again, you could do it by tracing.

Translations

When a shape is translated all its points move the same distance in the same direction. Again, the image and the original shape are congruent. For example this shape ▷ could translate to this ▷ .

To describe a translation, you need to say how far it moves across, left or right, and how far it moves up or down.

Exam tip

A common error is to move the wrong number of squares. Check by taking one point on the original shape and carefully counting to the corresponding point on the image (the translated shape).

EXAMPLE 3

Translate the shape four squares to the right and two squares down.

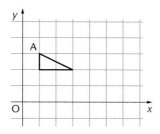

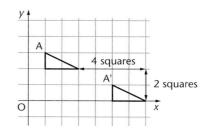

Point A has moved 4 along and 2 down and the other corners have moved in exactly the same way.

Describing transformations

As well as carrying out transformations you also need to be able to describe a transformation that has been done.

If it is a reflection you need to describe the mirror line.

If it is a translation you need to say how far the shape has moved, and in what direction.

EXAMPLE 4

For each of these diagrams describe fully the transformation of T to T'.

a)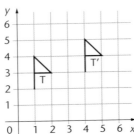

b)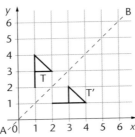

a) The image is the same way round and the same way up as the original shape, so it is a translation.

It has moved 3 squares to the right and 1 up. Check by counting squares.

This is a translation of three squares to the right and one up.

b) The image is the opposite way round from the original shape, so it is a reflection.

To find the mirror line, draw lines between matching points in the original shape and the image.

Then draw a line through the midpoints of these joining lines.

Then label the line or give its equation.

This is a reflection in the line AB or $y = x$.

Creating patterns

Successive transformations can be made on a shape, either in a line or around a point, to create patterns.

EXAMPLE 5

Reflect the triangle in the line $y = 0$ then in $y = -x$, $x = 0$, $Y = x$ and so on to make the full pattern.

T_1 is the reflection of T in $y = 0$, T_2 is the reflection of T_1 in $y = -x$ and so on.

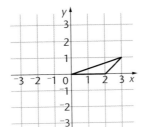 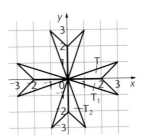

EXERCISE 20.1A

Use squared paper to answer these questions.

1 Reflect this shape in the line AB.

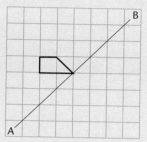

2 Rotate this shape through a half-turn about the origin.

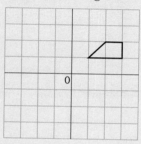

3 Translate this shape two squares to the left and one square down.

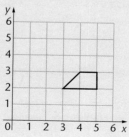

4 Reflect this shape in the line AB.

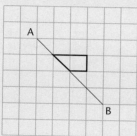

5 Rotate this shape through a three-quarter-turn anticlockwise about the origin.

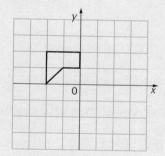

6 Translate this shape three squares to the right and two squares up.

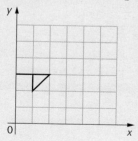

Exercise 20.1A cont'd

7 Rotate this shape through a quarter-turn anticlockwise about is centre A.

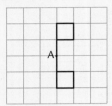

8 Reflect this shape in the line AB.

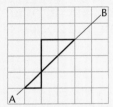

9 Reflect the shape in the line AB, then in a vertical line two squares to the right of AB, and so on to give a strip pattern. Show at least eight shapes.

10 In each of these diagrams, describe fully the transformation that maps A onto B.

a)

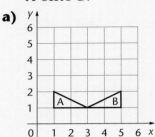

b)

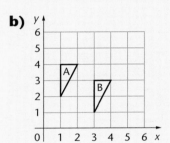

EXERCISE 20.1B

Use squared paper to answer these questions.

1 Reflect this shape in the line AB.

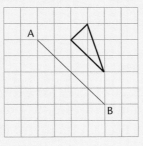

2 Rotate this shape through a quarter-turn anticlockwise about A.

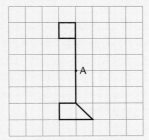

3 Translate this shape five squares to the right.

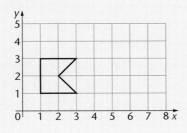

4 Reflect the shape in the line AB.

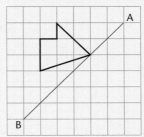

5 Describe the transformation that maps A onto B.

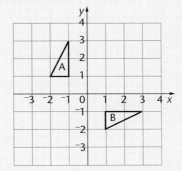

6 Rotate this shape through a quarter-turn anticlockwise about the origin.

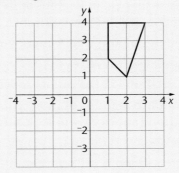

Exercise 20.1B cont'd

7 Translate this shape two squares to the left and four squares up.

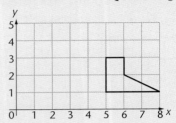

8 Reflect this shape in the line AB.

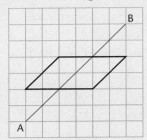

9 Draw a set of axes for *x* and *y* and label them both from ⁻4 to 4.

   a) Plot the points (⁻2, 1), (⁻1, 1), (⁻1, 3), then join them up to make a triangle. Label it A.

   b) Rotate A through a half-turn about the origin and label the image B.

   c) Reflect A in the *y*-axis and label the image C.

   d) Describe fully the transformation that maps B onto C.

10 Draw a set of axes for *x* and *y* and label them both from ⁻4 to 4.

   a) Plot the points (0, 0), (1, 3), (1, 1), then join them to make a triangle.

   b) Reflect the triangle in the line *y* = *x*.

   c) Rotate the shape made by the two triangles by a quarter-turn anticlockwise about the origin.

   d) Reflect all the shapes you now have in the *x*-axis to complete the pattern.

Key ideas

● When a shape is reflected, each point on the shape and the corresponding point on its image are the same distance from the mirror line, but on opposite sides.

● When describing a reflection, you need to define the mirror line.

● A translation moves every point on the shape the same distance, in the same direction.

● In a rotation, all the points move through the same angle about the same centre.

● When describing a rotation, you need to identify the centre, the angle and the direction (anticlockwise is positive).

● A translation is defined by movement to the left or right and up or down.

● When a shape is reflected, rotated or translated, it stays exactly the same shape. The shape and its image are congruent.

● When describing the symmetry of a shape, both line symmetry and rotational symmetry need to be considered

Compound measures

21

Compound units

Some measures depend on others, which means you need to multiply or divide other measures.

One important example of this is:

$$\text{average speed} = \frac{\text{total distance travelled}}{\text{total time taken}}$$

The units of your answer will depend on the units you begin with.

EXAMPLE 1

Find the average speed of an athlete who runs 100 m in 20 s.

Average speed = $\frac{100\,\text{m}}{20\,\text{s}}$ = 5 m/s.

EXAMPLE 2

Find the average speed of a delivery driver who travelled 45 km in 30 minutes.

The average speed = $\frac{45\,\text{km}}{30\,\text{minutes}}$ = 1·5 km/minute

However, the speed here is more likely to be needed in kilometres per hour. To find this, first change the time into hours.

So the average speed = $\frac{45\,\text{km}}{0\cdot5\,\text{h}}$ = 90 km/h.

You may also be able to see other ways of obtaining this result.

Exam tip

The units tell you which way round to divide so density measured in g/cm³ means grams (mass) divided by cm³ (volume).

Other examples of compound units are:

- density = $\dfrac{\text{mass}}{\text{volume}}$ with units such as g/cm^3.

- population density = $\dfrac{\text{population}}{\text{area}}$ with units such as number of people/km^2.

EXERCISE 21.1A

1 Find the average speed of a train which covers 120 miles in one and a half hours.

2 Find the average speed of a runner who covers 200 metres in 45 seconds.

3 Calculate the density of a metal of mass 450 gm and volume 40 cm^3.

4 A town has a population of 70 000 in an area of 12 square kilometres. Calculate its population density, (to the nearest whole number).

5 A motorbike travels 5 miles in 4 minutes. Calculate its average speed in miles per hour.

6 A bus travels at 7 m/sec on average. How many kilometres per hour is this?

7 A block of plastic with volume 25 cm^3 has a density of 0·6 gm/cm^3. What is its mass?

8 A town has a population of 120 000; its population density is 9500 people per square mile. What is the area of the town?

9 A runner's average speed in a 75 metre sprint is 8 m/s. Find the time he takes for the race, to the nearest 0·1 sec.

10 A car travels 120 km in 96 minutes. What is the average speed in km/hr?

11 A motorist leaves Derby at 10:55 and arrives in Leicester $1\frac{1}{4}$ hours later. The distance from Derby to Leicester is 50 miles.

  a) At what time did he arrive in Leicester?

  b) What was his average speed?

12 A boat leaves harbour at 0235 and arrives at its destination at 0405. It travelled at an average speed of 9 miles per hour. How far did it sail?

EXERCISE 21.1B

1 Find the average speed of a plane which travels 2250 miles in 5 hours.

2 Find the average speed of a boat which sails 140 km in 9 hours.

3 Calculate the density of a block of wood of mass 750 gm and volume 80 cm^3.

4 Smeeton has a population of 180 000 in an area of 20 square kilometres. Calculate its population density.

5 A train covers 0·6 miles in 1 minute. Calculate its average speed in miles per hour.

6 A cyclist travels at 6 m/s on average. How many kilometres per hour is this?

7 A rubber ball has a volume of 35 cm^3 and a density of 0·8 gm/cm^3. Calculate its mass.

8 A small island has a population of 12 360. The population density is 24 people per square mile. What is the area of the island?

9 A runner's average speed in a 200 m race is 4·8 m/s. Find the time she takes for the race, to the nearest 0·1 sec.

10 A car travels 20 km in 12 minutes. What is the average speed, in km/hr?

11 Fred drives for 5 miles through a town and 16 miles on a motorway. The time taken to go through the town was 30 minutes and he travelled for 15 minutes on the motorway.

What was his average speed for the whole journey?

When estimating measures in unfamiliar contexts, try to compare them with measures you do know.

Discrete measures can be counted. They can only take particular values.

Continuous measures include length, time, mass, etc. They cannot be measured exactly.

Revision exercise

In all of these questions make sure you state the units of your answer.

1 A triangle has vertices at A(6, 2), B(6, 7) and C(3, 5). Draw triangle ABC on squared paper and calculate its area.

2 A pentagon has vertices at A(2, 1), B(6, 1), C(6, 3), D(5, 6) and E(2, 3). Draw the pentagon on squared paper and calculate the area of

  a) the rectangle ABCE

  b) the triangle EDC

  c) the pentagon ABCDE.

3 Calculate the area of each of these parallelograms.

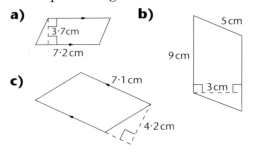

a)   3·7cm   7·2cm

b)   5 cm   9 cm   3 cm

c)   7·1 cm   4·2cm

4 Calculate the area of each trapezium.

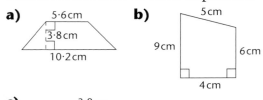

a)   5·6cm   3·8cm   10·2cm

b)   5 cm   9cm   6cm   4cm

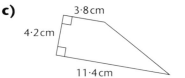

c)   3·8cm   4·2cm   11·4cm

5 A room measures 4·2 m long by 3·2 m wide by 2·6 m high.

  a) Calculate the volume of the room.

  b) Calculate the area of the four walls and the ceiling.

Jane paints the walls and ceiling.

1 litre of paint covers 13 m². The total area of the windows and door is 5 m².

  c) How much paint will she need?

6 Amy has counted the number of matches in ten boxes. Here are the results.

80	83	85	84	86
82	84	85	85	86

  a) Make a frequency table for Amy's results.

  b) What is the mode of her results?

  c) Use the frequency table to calculate the mean number of matches.

  d) Use the original results to calculate the mean as a check.

7 A market researcher has counted the number of people living in the houses on North Close.

No. of people	1	2	3	4	5	6	7
No. of houses	4	6	9	12	7	3	1

  a) How many houses are there?

  b) What is the modal number of people?

  c) What is the mean number of people? (To 1 decimal place.)

Use squared paper for **8–13**.

8 Enlarge this shape by scale factor 2, centre the origin.

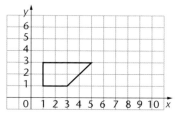

9 Enlarge this shape by scale factor 3, centre the point (3, 1).

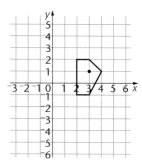

10 Reflect this shape in the line AB.

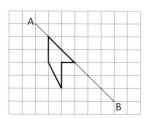

11 Translate this shape four units to the right and three units down.

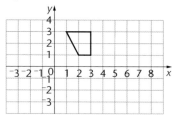

12 Rotate this shape through a three-quarter-turn anticlockwise about the origin.

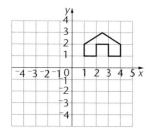

13 Describe fully the transformation that maps

    a) A onto B   **b)**   B onto D

    c) B onto C   **d)**   D onto E

    e) F onto B   **f)**   A onto G

    g) G onto B.

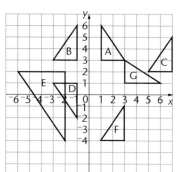

14 Sasha runs a 100 m race in 13·58 s. Calculate her average speed. Give your answer to a suitable degree of accuracy.

15 A cyclist travels 5 km in 20 minutes. Calculate her speed in kilometres per hour.

16 A metal weight has mass 200 g and density 25 g/cm^3. What is its volume?